DEFRAMING STRATEGY

How **Digital Technologies** are Transforming **Businesses** and **Organizations**, and How We Can Cope with It

DEFRAMING STRATEGY

How **Digital Technologies** are Transforming **Businesses** and **Organizations**, and How We Can Cope with It

Soichiro Takagi

University of Tokyo, Japan

World Scientific

NEW JERSEY • LONDON • SINGAPORE • BEIJING • SHANGHAI • HONG KONG • TAIPEI • CHENNAI • TOKYO

Published by

World Scientific Publishing Co. Pte. Ltd.

5 Toh Tuck Link, Singapore 596224

USA office: 27 Warren Street, Suite 401-402, Hackensack, NJ 07601

UK office: 57 Shelton Street, Covent Garden, London WC2H 9HE

Library of Congress Cataloging-in-Publication Data
Names: Takagi, Soichiro, author.
Title: Deframing strategy : how digital technologies are transforming businesses and organizations, and how we can cope with it / Soichiro Takagi, University of Tokyo, Japan.
Description: New Jersey : World Scientific, [2022] | Includes bibliographical references and index.
Identifiers: LCCN 2021034617 | ISBN 9789811243691 (hardcover) | ISBN 9789811243707 (ebook) | ISBN 9789811243714 (ebook other)
Subjects: LCSH: Business--Technological innovations. | Organizational change--Management. | Technological innovation--Economic aspects. | Disruptive technologies.
Classification: LCC HD45 .T349 2022 | DDC 658.4/062--dc23
LC record available at https://lccn.loc.gov/2021034617

British Library Cataloguing-in-Publication Data
A catalogue record for this book is available from the British Library.

For any available supplementary material, please visit
https://www.worldscientific.com/worldscibooks/10.1142/12453#t=suppl

Desk Editors: Balamurugan Rajendran/Sandhya Venkatesh

Typeset by Stallion Press
Email: enquiries@stallionpress.com

Printed in Singapore

About the Author

Soichiro Takagi is an Associate Professor at Interfaculty Initiative in Information Studies at The University of Tokyo. He also serves as a fellow at the Art Center of The University of Tokyo and a senior research fellow at Center for Global Communications (GLOCOM) at International University of Japan (IUJ). Through his career, he served as a Professor and the Director of Blockchain Economic Research Lab at GLOCOM, and an Asia Program Fellow at Harvard Kennedy School, etc.

His major field is information economics and digital economy, focusing on the relationship among information technology, organizations, and the economy. He has examined a variety of topics including offshore outsourcing, cloud computing, open data, sharing economy, digital currency, blockchain, and digital platforms. He has authored many books and articles, including *Reweaving the Economy: How IT Affects the Borders of Country and Organization* (University of Tokyo Press) and *Blockchain Economics: Implications of Distributed Ledgers* (World Scientific, co-authored). He received the KDDI Foundation Award in 2019. He received a Ph.D. in information studies from The University of Tokyo.

Acknowledgments

The author thanks all the following experts for inspiring discussions in the course of writing this book. I obtained precious knowledge on the development of platform businesses and the strategies of the software industry from Professor Marshall Van Alstyne at Boston University, and Professor Michael A. Cusumano at the Massachusetts Institute of Technology. I gained knowledge on the latest development of IT business in China from Associate Professor Asei Ito at The University of Tokyo, and Mr. Takeshi Yamaya, an IT journalist covering Asia, as well as Ms. Weilin Zhao, a senior researcher at the Itochu Research Institute. I owe my knowledge on the development of technological architecture to Mr. Takuya Oikawa, CEO of Tably, Co., and great insights on value creation and circulation in local communities to Professor Hideyuki Tanaka at The University of Tokyo and Chief Digital Officer Yasushi Fujii at Nishiaizu Town. I also thank Ms. Mayumi Morinaga at Hakuhodo DY Media Partners for their valuable comments on influencer marketing. I acknowledge that members of GLOCOM at the International University of Japan contributed to the contents not only by providing information but also by giving me valuable comments on the contents of the book. I thank Mr. Yasuo Kyobe at Shoeisha for his acceptance of the publication as well as valuable comments on the contents for the original Japanese version. My sincere gratitude goes to Ms. Amirova Nargiza, Ms. Venkatesh Sandya, and Mr. Balamurugan Rajendran of World Scientific group for the decision of publication and editorial works for the English version. A part of the book is based on research supported by JSPS Kakenhi JP15K00460. I thank all the people involved for their support and cooperation.

Contents

Introduction

The aim of this book is to reveal the fundamental impact of digitalization on businesses and the economy. As a key to understanding this fundamental impact, this book proposes the concept of "Deframing."

Deframing is a framework for understanding the social and economic impact of various digital technologies. It is also a basic force to transform a wide range of economic aspects such as business models, corporate strategy, workstyle, career design, and how to learn. Deframing is a made-up word, which represents the collapse of existing frames, and is an indispensable viewpoint to foresee how society will change in the coming future.

Deframing illustrates and explains phenomena in which internal elements in existing frames of products, services, and organizations are dissolved and reintegrated beyond traditional frames, thus enabling the provision of optimized services for others. In other words, existing frames of services and organizations are disappearing in the deframing age. Products and services are transformed from a packaged form designed to satisfy everybody's demand moderately, to more specifically optimized ones for each user. It applies not only to products but also business organizations. From the workers' perspective, an increasing number of people conduct business as individuals by taking advantage of their skills and resources, instead of working as an employee of existing frames of companies.

These phenomena are already observed universally in society. For example, when you buy your shoes, it is possible to design and customize as you like and order them instead of browsing through many stores looking for shoes that fit your preferences. It is also possible to make hardware with a small number of units based on your preferences and designs. In other cases, social networking services are transforming themselves from simple communication platforms to platforms where individuals can advertise as influencers and sell products, replacing traditional channels such as advertising media and retail stores. Using crowdsourcing platforms, individuals are taking over the business that was previously conducted only by professional firms, such as in the software and design sectors.

Although the term "Digital Transformation" (DX) has received attention in recent years, until now, the term suggested nothing more specific than that digital technology will transform business and services at a deeper level. Neither a discipline or theory of digital transformation, nor its impact on society, have been clarified. However, continuously evolving technologies are exerting deep impacts on the structure and function of society the beyond mere improvement of efficiency. This book reveals how DX affects society and the economy. This book is based on the deframing concept and shows its theoretical background and, how it is applied to understand the latest developments in the economy with specific examples. In addition, it also proposes how companies and individuals should respond to changes in the age of deframing.

I have participated in the discussion on the implications of the latest digital services and its social and economic impact at the Interfaculty Initiative in Information Studies at The University of Tokyo and the Center for Global Communications (GLOCOM) at the International University of Japan. To conduct research on the latest digital services, in addition to my base location, Tokyo, I visited various places such as New York, Silicon Valley, Shanghai, Hangzhou, Shenzhen, and London, where I am serving as a research fellow at the Center for Blockchain Technologies at University College London. In the course of research on digital services and those impact on society, the concept of deframing was elaborated as the one that can best explain the social and economic transformation driven by digital technologies, based on economics perspective. It would be my sincere pleasure if this book and the concept of deframing help the readers foresee the impact of technological evolution on society in the future.

This book is constructed by three parts. Part I outlines the concept of deframing with its background. Chapter 1 introduces the concept of deframing, which consists of three elements: dissolution and reintegration, specific-optimization, and individualization. This chapter briefly explains the idea of the three elements.

Chapter 2 explains the theoretical background of deframing to show why modern technological evolution causes social and economic change, which is characterized by the concept of deframing. Deframing is deeply related to the structure of corporate organizations, that has been extensively studied in organizational economics. This chapter examines why frameworks of corporate organizations have been necessary throughout history and how they are changing with the development of information technology, and why these changes are causing deframing. Based on the examination, this chapter shows that the economy is shifting from "the age of large hierarchical corporations" to "the age of new personal industry" in a historical viewpoint. If you are interested more in specific cases than its theoretical background, you can skip Chapter 2 and directly proceed to Chapter 3.

For Part II, the process of deframing consists of three chapters that correspond to deframing's three elements. Chapter 3 dives into the first element, "Dissolution and Reintegration." It examines how existing business models are dissolved by digital technology, and how they are restructured as new business domains. This chapter discusses the strategy and background of dissolution and reintegration, with a variety of examples around the world. This chapter also suggests possible strategies to respond to the trend of dissolution and reintegration.

Chapter 4 focuses on the second element, "Specific-optimization." This chapter argues that the age of ready-made, "one size fits all" products and services has come to an end, and explains why it is becoming important to provide values that are optimized for specific needs of customers and users, and how this became possible through technological innovation. Specific-optimization is an important viewpoint not only for businesses but also for public and regional management. These social issues are also discussed in the latter half of the chapter.

Chapter 5 discusses "Individualization," the third element. Since digital technologies drastically improved access to information, even individuals can provide services that are comparable to traditional organizations. Based on this background, the number of freelancers is increasing globally. Platform technologies, such as those seen in the sharing economy, are a system for matching and mediating trust that enables individuals to provide services replacing corporate organizations. On the other hand, the importance of communities that support individualized service providers is growing, as seen in the global increase of co-working spaces. This chapter focuses on the background and challenges of individualization and provides an analysis of co-working spaces that have the potential to provide competing function replacing corporate organizations.

Part III discusses the strategy for organizations and individuals in the Deframing Era. It focuses on the issue of collaboration between people, the career strategy of individuals, and the future social challenges in the deframing age. Chapter 6 discusses the issues of trust, that is one of the most important elements that enables deframing, and the key to the success of modern platforms services. In a digitalized society, trusting others to enable commerce and business transactions between unknown individuals has been a long-time challenge. This chapter also discusses how blockchain technology, sometimes known as "the Internet of trust," is related to trust in the business context.

Chapter 7 focuses on the personal strategy for individuals in the deframing age. Super-aging populations are causing serious problems in social systems globally. For example, social security systems are in financially critical situations, and "lifetime employment" is virtually becoming an illusion for most companies whose sustainability would not necessarily be longer than a human life. When individualization of industry is accelerated in an aging society, it is important to ask how we should learn, and how we should construct our careers. This chapter explores individual strategies that each of us should consider in the deframing age.

Chapter 8 presents the challenges and prospects of deframing. The deframing age is the time of individualization and of the emergence of gigantic platform providers that mediate transactions of those individuals. This chapter addresses the issue of oligopoly of platforms and the accumulation of personal information. It also discusses issues such as privacy, and the potential risk for individuals to feel isolated from each other during this era of individualization. This chapter provides remarks on these topics and proposes possible solutions. In addition, it also discusses strategies for cities to promote innovation in the deframing age.

This book provides the principle of understanding the social impact of rapidly evolving information technologies and to consider responses to changes. I will be happy if this book can offer you a new way to think about business, career, and public policy for the coming future.

Part I

Introduction of Deframing

Chapter 1

The End of "Economies of Scale"

The 20th-century economy was characterized by economies of scale, a fundamental concept in economics. Whether it involved products or services, and regardless of each user's demand, the 20th-century economy achieved economies of scale by packaging consumer demands into uniform products, and distributing them on a massive scale. This was a result of the Industrial Revolution in the 18th century, which transformed the economy from the one depending on craftmanship and small-scale production to the one of large-scale production of same products in a factory, supported by massive distribution through railway networks and steamships.

Since then, the economy has been based on the same industrial concepts to pursue profits, which is applicable to the Ford model T and Toyota production system as well as Microsoft's Windows operating system and Apple Macintosh computers. It is based on the idea of focusing on packaging the functions that satisfy the majority of the customers and avoiding customization while selling it on a large scale globally. This has been the golden rule of the economy for a long time.

Certainly, there are several benefits of selling packaged products. On the producers' side, it is possible to reduce production costs by producing the goods or services based on the same design. For consumers, it is possible to find goods or services that are moderately satisfactory without addressing their most specific preferences. However, goods or services that are packaged to a certain "frame"' actually include a massive amount of waste. For example, your smartphone would contain applications that you have never used or even seen since you bought the phone. You may think that it is OK because it is free, but its development cost is certainly included in your purchase. Or, you must simply be content to use the goods that are not actually what you want but rather just a relatively good substitute.

Conversely, not only products or services but also we, as workers, have hoped and chosen to be packaged in a certain "frame" and take actions to fit into it. We accepted packaging ourselves to be a graduate from a certain university by enrolling in prestigious universities to become a banker or a lawyer, and even define ourselves as "corporate employees" when we start working. Each person should have various talents, skills, advantages, and disadvantages, but it is not efficient if we start considering these varieties. Therefore, we chose to produce ourselves through frames such as "I am [a certain] university graduate" to show that we have certain skills. Then, we become an employee of a certain company and have defined our career life with the company's image. However, individuals have their own strengths and weaknesses among the unlimited variety of attributes and dimensions and have differing resources such as free time and skills. In the past, we have worked within a "frame" of a company, so we could not fully utilize those resources.

Recent advances in technology have the potential to overcome the inefficiencies of such packaging. It is now much easier to take individual elements inside a packaged frame, and match, combine, customize them. It has become possible to combine various values inside providers with various demands inside users more closely, and to mediate transactions.

When we look at the economies of scale by emphasizing packaging from the standard of modern technological capabilities, it becomes clear that it is wasteful and obsolete, and rather the combination of individual elements inside packages became more important. In this book, I propose the concept of deframing that meets users' needs by dissolving previously packaged products, services, and organizations into fundamental elements, and then flexibly reorganizing them. I will elucidate the essence of this concept in light of concrete examples, and consider how companies and individuals should work to transition away from economies of scale. This will be called a "Deframing strategy."

Definition of Deframing and Its Three Factors

First, I would like to define deframing. Deframing refers to providing services that meet user needs by combining and customizing internal elements beyond traditional frameworks of services and organizations. If you look at online IT service companies that are rapidly growing, such as Google, Amazon, Line, Mercari, Alibaba, and Tencent, you can see that their growth strategy is in line with this principle of deframing. The concept of deframing was born as a framework that can best explain the social and economic transformation driven by innovative services and business models around the world, thereby providing a perspective on future changes. Deframing is a fundamental principle in modern digitalized services, and it is worth for companies to rethink their business models based on the concept of deframing.

Deframing is largely composed of three elements: "dissolution and reintegration," "specific-optimization," and "individualization." We will briefly examine these three elements in Figure 1-1.

"Dissolution and reintegration" mean that service providers deconstruct existing frameworks such as business domains, business models, and business processes; extract internal elements; and reintegrate them across traditional boundaries. It causes the transformation and redefinition of business domains because it shuffles and reintegrates elements that used to belong to different business domains. "Specific-optimization" customizes and personalizes products or services to meet each user's demand beyond packaged forms. It leads to the transformation of business models that used to focus on economies of scale by providing uniform products or services to customers. "Individualization" suggests an economy where individuals play a more important role as service providers rather than working as employees of hierarchical organizations. It is the

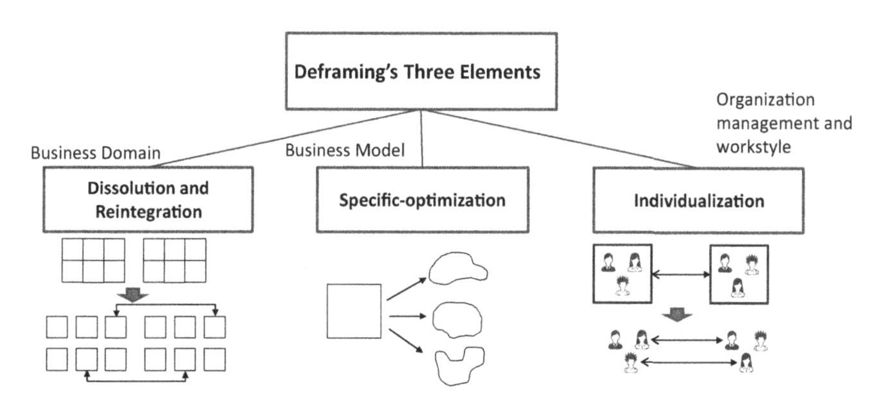

Figure 1-1 Structure of deframing

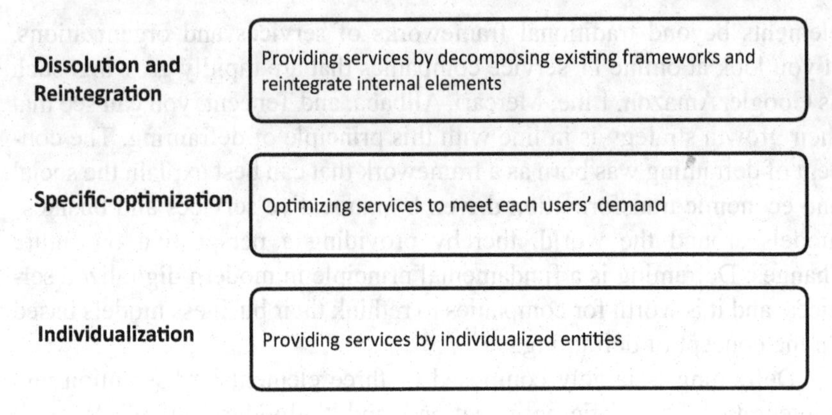

Figure 1-2 Three factors of deframing

transformation of the current workstyle and employment pattern to a new one that emphasizes more autonomy and flexibility.

These three elements might seem to belong to quite different aspects of the economy, but as we will see in Chapter 2, they are caused by the same mechanisms and forces, just surfaced in different aspects of the economy (Figure 1-2).

Dissolving and Reintegrating Elements Beyond Frames

Let us look at the first element, "dissolution and reintegration," where elements included in the incumbent products or services are disassembled and reintegrated. Traditional businesses and services are often provided as "packages" by combining several elements. An example is a university, which is considered as a packaged service composed of a wide range of elements such as classes, conferring degrees, guiding students to improve their abilities, a community function for students, a social function to spend extracurricular time, etc. In other examples in the hotel industry, it is a packaged service comprising room spaces, cleaning services, food and beverage services, concierge services, and in some cases, hot springs (particularly in Japan). Similarly, department stores consist of various services such as the selection of products by connoisseurs, purchase support by reception, and payment functions at cash registers. At training gyms, training equipment, pools, training advice and guidance, and showers are packaged.

Not all these packaged elements are used equally. Some college students only need classes, while others desire club communities. Some people in gyms may also want a good coach to teach, but do not need heavy weight training equipment. It is now possible to match available values and people who need them in more micro-scale units using the advancement of information technology. Therefore, the "frames" to package various elements together are no longer necessary. Mr. Ken Suzuki, CEO of Smart News, in his book, describes this as "smooth society," which "is realized after the 'membrane' was removed and the 'nucleus' was connected."[1] This means that distinctions that were separated and defined by "membranes" disappeared, creating a world in which only the contents were connected as needed.

The packaging of various elements into a uniform product is a good strategy to meet the demand of most users, but when we examine the package in detail, we can find unnecessary elements. In the "deframing age," it is important to respond to individual needs and to subdivide the value that used to be provided in packaged services. For example, Google has removed the boundaries of traditional access methods to information such as books and newspapers and provided access directly to the information that you really need. Apple's iTunes has made it possible to purchase only the songs that you want to listen to, instead of using a package of CD that couples many songs. In addition, streaming and subscription-based services enabled you to freely access to specific songs that you want to listen to.

In the financial sector, conventional banks provide comprehensive services such as retail stores, ATM networks, payment functions, and loans. However, PayPal has focused on the function of transferring money between users. By narrowing down the functions and focusing on the convenience that is suited to the Internet age, it has become possible to provide remittance services around the world at extremely low costs. It now manages over 392 million accounts in the first quarter of 2021.[2]

In this way, this book considers above-mentioned services as a dissolution that provides services by extracting the elements for which you have an advantage, that has the greatest demands, and that has the greatest challenges out of incumbent packaged services. Then, it is possible to combine the extracted elements and build a new service to pursue the

---

[1] Suzuki, Ken. (2013). *Smooth Society and Its Enemies*: Keiso Shobo (original in Japanese, translated by the author).

[2] *Source*: Statista, https://www.statista.com/statistics/218493/paypals-total-active-registered-accounts-from-2010/ (Accessed on July 15, 2021).

"economy of scope." At this time, the combination across completely different sectors can occur. Communication functions in the telecom sector and remittance functions in the financial sector can be combined, and a marketplace service as infrastructure can be integrated with credit information services. These are already occurring throughout the world. The process from dissolution to reintegration, with expansion of scopes, is explained in detail in Chapter 3.

Scaling Specific-Optimization

The second element of deframing is to provide a service that is specifically optimized for each user, avoiding uniform products or services for all users. In traditional management practices, specific-optimization is difficult because it costs too much, particularly when the service is provided to many users. Even if the difference between person A's and person B's needs is subtle, it is costly to customize for individual demands, therefore we have packaged and offered the "common divisor" services that everyone was reasonably satisfied.

However, recent technological advances have made it possible to meet individual demands without increasing cost. One such concept is mass customization. Nike's service, called NIKE BY YOU (formerly NIKE iD), has made it possible to manufacture shoes with designs tailored to each user's preferences. This service became possible by integrating Web interface design that converts user needs into formalized data, and production technology that immediately transmits them to the factory and implements them in production, and supply chain technology that smoothly delivers them to users.

Similarly, information services are tailored for each customer. For example, the Amazon.com website has a completely different appearance for each user. Based on past purchasing history, a lineup of products that are likely to interest you would be displayed. If you log out of Amazon and try to access it again, you will see the difference from your usual experience. At Google, you will also find search results that are different for other users. It is tailored to return the best search results for the user by analyzing past preferences and search results. These services are used by hundreds of millions of people, but it is possible to customize for individual needs. Customization of such information services is called "personalization."

In recent years, such personalization has become possible by using machine learning, an artificial intelligence (AI) technology. By analyzing the past results and experiences of each customer, it became possible to fine-tune the algorithms and parameters, and then personalize the services to fit the user demand. Behind the website, servers can analyze where you looked at the search results, where you spent your time in the website, and which products you gave serious consideration. By analyzing this information, it is possible to provide the correct features for each person. In this way, even if the provided value is customized to each customer, no additional cost is needed once the mechanism is built as a software. The so-called "marginal costs" are diminished to nearly zero,[3] minimizing the additional cost of customizing services.

These specific-optimizations extend to not only software and Internet services but also to hardware. With the advent of 3D printers, if you have design data, you can create the design with your nearest 3D printer. If you want to change the design, you can modify it by yourself or add some modules. The emergence of people called "Makers," who make their own hardware, symbolizes the tailor-made hardware customized for each user. The festival for makers, "Maker Faire" attracts 200,000 people annually to its major events in the US and provides occasions for more people to gather in its independently produced "Mini Maker Faires" globally.[4]

The role of "platforms" in providing specific-optimized services is also important. To meet the diverse needs of users, it is effective to have a wide range of service variations. Some book buyers may want not only books but also sweets and drinks, while others may want to buy a computer along with books. To meet each different need, suppliers need to pursue the "economy of scope." With the power of the Internet, it is possible to flexibly combine various services and third-party providers to create services that are dynamically aligned for each customer. Digital platforms are what makes this possible.

Many books, articles, and reports discuss the definition of platform, but based on the literature review, I define it as follows: "a digital platform is a service that connects different groups of entities and mediates transactions through digitalized mechanisms to take advantage of network

[3]Rifkin, J. (2014). *The Zero Marginal Cost Society*, St. Martin's Griffin.

[4]Maker Faire A Bit of History. https://makerfaire.com/makerfairehistory/.

externality."[5] As will be seen frequently in the following sections, platforms play a significant role to meet the demand of customers. With the development of mass customization technology and platforms, it is becoming possible to create a system that scales globally while still trying to optimize services for specific demands.

Individualizing Industrial Structure

The third element is the "individualization" of service providers. In the past, service providers have belonged to large organizations such as companies and worked in collaboration with other employees within the company. This is because it is easier to work with people who share the same cultural norms and institutional systems within an organization. Without organizations, people had to elaborate on their proper counterparts and evaluate them cautiously so that they are not cheated by unknown collaborators.

However, with the evolution of technology, particularly in distributed trust systems enabled by platform mechanisms, individuals are able to easily cooperate with people without relying on the organizations' trust and system of coordination. In other words, the first element, the "dissolution and reintegration" principle is applied to how we work and how we collaborate with others. These phenomena are clearly observed in the increase of the sharing economy, freelancing, and side jobs, and having multiple jobs. In Japan, 85.3% of companies did not permit employees to take side jobs in 2014, but the percentage of employees who want to take part in side jobs increased from 4.4% in 1992 to 6.4% in 2017.[6] In particular, the gap between the willingness and actual conduct of side jobs is wider in workers with regular employment than those with non-regular employment, and 5.4% of regular employment workers wish to take part in side jobs whereas only 2.0% of them have realized it in 2017, and the gap is widening.[7] The Japanese government lifted the employment

---

[5]Takagi, S. (2020). Literature survey on the economic impact of digital platforms. *The International Journal of Economic Policy Studies* 14: 449–464. https://doi.org/10.1007/s42495-020-00043-0.

[6]Ministry of Health, Labour and Welfare. https://www.mhlw.go.jp/stf/shingi2/0000179566.html, https://www.mhlw.go.jp/file/05-Shingikai-11201000-Roudoukijunkyoku-Soumuka/0000179562.pdf.

[7]Statistics Bureau of Japan. https://www.stat.go.jp/data/shugyou/2017/pdf/kgaiyou.pdf.

contract model restrictive to side jobs in January 2018, responding to the demand of workers and social change.

Another factor enabling individualization is the fact that technology enables anyone to access cutting-edge information and knowledge and to expand their capability to conduct business on their own. Compared to the time when you needed a typist to write a letter to others, now you can write e-mails, software codes, or even launch smartphone apps by yourself. The need to divide and share tasks with others is decreasing. In addition, the Internet has enabled anyone who is not at the top of hierarchical organizations to access the latest information freely and become knowledgeable on a certain topic. In some cases, people at the bottom of the corporate hierarchy might be more knowledgeable than senior managers or even CEOs in specific fields. These changes suggest that individuals are empowered by the democratization of information driven by the Internet.

In this context, there is a greater chance to harness the potential of individuals. In the past, we have defined our jobs in frames of companies and organizations such as bankers, public servants, and university professors, while we had to change our skills continuously to meet the demands of the organizations. However, in the future, we must compete with expert individuals who operate outside of the organizations and specialized in similar tasks. If we settle in a frame of organizations and wait for orders from the company, we may be replaced by outside freelancers. Therefore, we need to develop our strengths to pursue what we are good at and what we can persistently excel at. As seen in crowdsourcing platforms, even if your strength is a niche ability, you can reach a certain number of consumers with needs if you provide it on a global scale. Recently in Japan, even salary-based workers have been allowed to hold side jobs. This is also based on economic rationality to take advantage of the various abilities of workers.

On the other hand, tasks that anybody can provide and that is difficult to be differentiated are required to be automated in the future. Even before the emergence of AI, processes with high demand and that are easily standardized tend to stimulate technological innovation and be replaced by computers. For example, in the case of transcription of meeting minutes, humans were transcribing the text by listening to the recording of the meeting in the past, but thanks to the technological development of speech recognition and natural language processing, accurate transcription can be easily achieved with a speech-to-text application. High-quality language translation can be conducted by machines such as with Google Translate.

In other words, the abilities of "the greatest common divisor" tend to be the object of automation by AI because they have a significant demand and have had a massive human resource cost. Therefore, the best strategy for individuals is to foster unique abilities that others cannot compete with and utilize them on a global scale. This is a kind of challenge toward the concept of "occupation." This is also the consequence of the deconstruction of "frames." On the other hand, the COVID-19 crisis that started in late 2019 revealed the vulnerability of freelancers and the self-employed. The challenges facing job vulnerability because of deframing will be further discussed in Chapter 7.

Efficient Resource Allocation as a Fundamental Force

The fundamental force behind deframing is the rapid development of information technologies. For instance, a sharing economy, which quickly developed throughout the world, can be regarded as a service that removes the frames of existing "companies" and "employees," and connects individuals with resources and potential customers. Airbnb, one of the representative firms of the sharing economy, is a platform used to help travelers seeking accommodation to rent empty rooms, apartments, and homes owned by individuals. The emergence of a sharing economy has revealed that these resources are available in society, even if they are not owned by the professional hotel industry. In addition, Airbnb cultivated the hidden demand that people want to communicate with local residents and also want to spend time like those who live there, not just staying in gorgeous hotels.

Uber also utilizes the same mechanisms to offer a taxi service based on the sharing economy concept. It showed that the availability of resources such as available cars, the ability of driving, and the time of drivers is not limited to the frames of "taxi company" and "taxi driver." Uber visualized such hidden resources and made it possible to gain value from them. The fundamental force behind these changes is efficient resource allocation in the concept of economics, where technological evolution enables the utilization of resources not currently in use and delivers them to the right place. In the past, there was a wall of trust between consumers and providers, making it difficult to transact between them. However, the growth of the sharing economy suggests that it transformed the core system of merchandizing transactions from "trust of corporate organizations" to "trust of individuals." Trust issues are discussed in Chapter 6 in detail.

In terms of resource allocation, the financial sector has long exhibited the importance of its mediation. For example, a loan allocates a financial resource from a place where the money is not in use to a place where it is in strong demand. Deposits are the inter-temporal allocation of financial resources from a time when the money is not in use for a time when it is really needed. Insurance allocates resources between those who are potentially at risk and those who are facing the realized damages.

Efficient resource allocation is transformed by various technological aspects. Blockchain technology and cryptocurrency have enabled anyone, not only central banks and authoritative organizations, to create a medium of exchange, a major function of money. These technologies provide a medium to record a variety of values that are produced in society, and to exchange them far more flexibly and in far smaller units. The more means of storing and exchanging values diverge, the more values that are captured and transacted become diverged. For example, electricity power generated in each house is recorded and transacted using blockchain technology, as explained in Chapter 3. The power of technology enables far smaller and more diverse units of resources to be allocated.

The above discussion to outline the concept of deframing is summarized in Figure 1-3. From a technological point of view, digital platforms

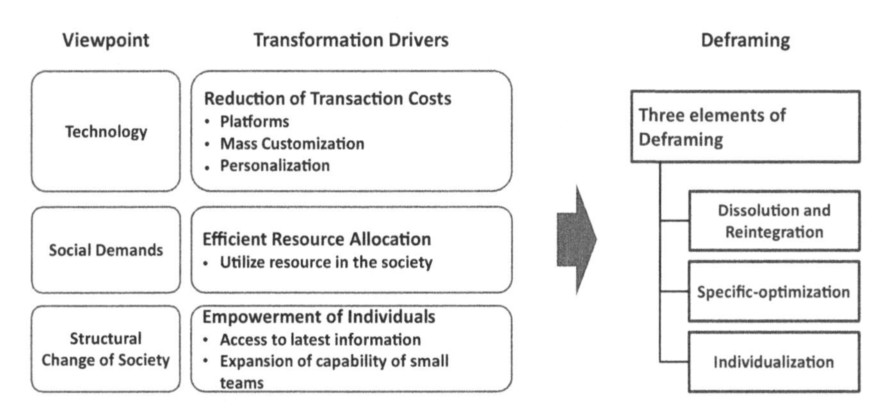

Figure 1-3 Outline of background of deframing

are the main driver of transformation. They have enabled the combination of resources in far more detailed units than before, and with far smaller transaction costs. In addition, mass customization and personalization technology reduced costs when businesses optimized their services for each customer. From the perspective of social demand, there is a continuous pressure on how we can utilize resources efficiently without waste. It is an eternal challenge in society to utilize idle resources and human skills to produce value. From the viewpoint of structural changes in society, individuals are empowered by the enhanced access to information and by harnessing their capability. These three transformation drivers are pushing our society to the age of deframing.

Deframing strategy is a method to deconstruct and reintegrate the elements that were previously packaged in a frame and optimize services for each customer, achieving scalability at the same time. It is also a strategy for individuals to harness their unique capabilities. Your customers are already getting used to the services that respond to their specific demands, so packaged services based on old models may be considered relatively expensive. Products that fit users, not users who fit products, are becoming the new standard. When people notice that they can obtain better outcomes by relying on individual experts specialized on a topic, they will not utilize the expensive services from companies. From an individual's perspective, it is becoming more important to design your career in the society where industrial structure is becoming individualized. Now is the time to take advantage of individual powers and collaborate across organizational boundaries.

Removing barriers, linking elements, and responding to individual needs — this is the basic strategy to survive in the 21st-century economy. In Chapter 2, based on the economics of information technology, we will closely examine why this is happening theoretically and how to interpret this change from a historical perspective.

Chapter 2
Deframing Mechanism

Deframing, which comprises "dissolution and reintegration," "specific-optimization," and "individualization," might seem to be complex and costly to implement compared to past economic patterns based on uniformity and packaging. Why is deframing economically rational compared to past economic systems? The answer lies in intertwined elements such as technological innovation, democratization of information, and the emergence of new business models. These elements change the structure of economic organizations and affect a wide range of aspects such as business models, social structure, and workstyles. This chapter reveals the theoretical background of deframing from the perspective of economics. You may skip this chapter and proceed to Chapter 3 if you are particularly interested in specific cases and strategies, but you will be able to understand the background and mechanisms of deframing at a deeper level by exploring this chapter.

Beginning of the End of Hierarchical Organizations

One of the key factors of the growing importance of deframing is end of the age at which the economy is based on hierarchical corporate organizations. Hierarchical organizations were a key driver of the labor-intensive economy in the 19th and 20th centuries. However, as the economy shifts from being labor-intensive to knowledge-intensive, the disadvantages of hierarchical organizations are becoming apparent. We will approach the

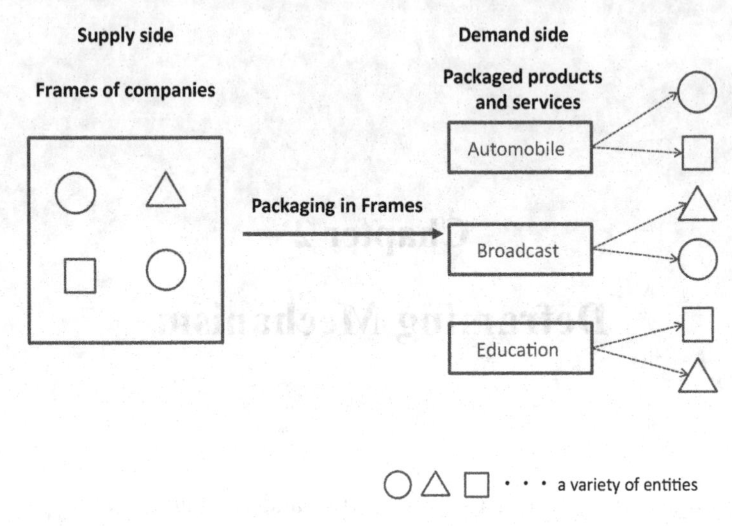

Figure 2-1 Past socio-economic system based on packaging

deframing mechanisms from the change in corporate organizations. Figure 2-1 shows the basic relationship between companies, products, and customers.

When we look at the supply side, the left side of the figure, "frames" of companies comprise various factors of individuals (various skills of workers) and other resources (facilities such as factories, data, software, natural resources, etc.). Most products and services in the market are provided through these frames of companies. Why then did we need the frames of companies at the beginning?

Why Were Corporate Organizations Required?

You may know the story of the pin factory of Adam Smith, the so-called father of economics. This story explains that to produce more pins, it is better to hire multiple employees who are specialized in a specific task and divide the work among them to produce pins, rather than one person performing all the tasks. Adam Smith showed that when one craftworker takes charge of all the processes to produce a pin, he or she can produce only one pin per day. However, if they divide work to many tasks of different craftworkers, such as the one who stretches the metal and the one who creates the head of pin, they can produce 4,800 pins per day per worker.[1]

This represents the idea that any individual or company can raise productivity by specializing in what they are good at and exchange those advantages. In addition to individuals and companies, even countries can benefit from producing what they are good at and exchange them as international trade rather than producing all goods or services in one country. The concept of "being good at something" is considered comparative advantage in economics. Entities do not have to be the best among all players in the economy in a certain task, but they are better off only if they focus on what they are good at compared to other things in a country, company, or among individuals. If entities specialize in their comparative advantage, their resources such as capital and skills for producing goods or services become sophisticated and accumulated, and they become even better at those tasks.[2] If we can efficiently promote cooperation among different entities who are specialized in a certain service, the whole economy can increase productivity and produce more with less resources.

There are roughly two measures to promote cooperation among independent entities: "the market," or "hierarchy."[3] In the market, different tasks or products are independently traded based on the price signals in a spot transaction. In the hierarchy, entities cooperate under the mechanisms of command and control backed up by authority. The issue of whether the market or hierarchy is better has been studied in organizational economics for many years.

Kenneth Arrow, the father of organizational economics and Nobel laureate for economics in 1972, stated that organizations are basically the mechanisms to achieve the benefits of collective actions when the price adjustment system does not function in the market, and are characterized

---

[1] Smith, A. (1776, 1982). *The Wealth of Nations*, Penguin Classics.

[2] Kenneth Arrow suggests that the benefits of collective action and having society are to "extend the domain of individual rationality" and "means by which individuals can more fully realize their individual values." In Arrow, K. J. (1974). *The Limits of Organization*, W. W. Norton & Company, Inc., New York, p. 16.

[3] The broadest meaning of an organization is any coordinated activities among individuals. See Barnard, C. I. (1971). *The Functions of the Executive: 30th Anniversary Edition*, Harvard University Press.

by authority as its foundation.[4] There are several reasons why the market may not function properly. For example, it is the case that there is a difficulty to describe the specifications and conditions for procurement in the market; therefore, uncertainty is too high for trading, and it is difficult to trade in the market when the trading counterpart does not behave with integrity.

Transaction Cost Theory Defining Organizational Form

These problems were deeply analyzed by economics scholars such as Ronald Coase (Nobel laureate in economics in 1991) and Oliver Williamson (Nobel laureate in economics in 2009). Nobel awards might seem to given to the topics that is far from daily life, but it has been given to near and dear topics such as "Why do companies exist?" One of the methods to explore this question is through the concept of "transaction cost." Transaction cost is the cost associated with the transaction itself, distinct from the cost of producing goods or services. When you trade goods or services, you have to pay costs associated with the transaction itself, in addition to the price of the good or service. In the case of buying goods in stores, travel to the store and the transport of purchased goods are considered transaction costs. Professor Chintagunta of the University of Chicago and co-authors quantified transaction costs associated with these purchases.[5] According to them, online shopping provides a value of 0.59 Euro per one kilometer travel, 0.69 Euro as time saving per 10 items bought, 0.53 Euro for picking up and putting 10 heavy/bulky items in a shopping cart, and 0.98 Euro for carrying 10 heavy/bulky items for one kilometer. Additionally, the cost of not being able to check the quality of perishables in an online store is 0.34 Euro per 10 perishables.

Similarly, transaction costs occur in the transaction of human tasks. If it costs less to transact in the market, this transaction is handled in the market, while if it costs more, it will be assigned to a person in

---

[4]Arrow, K. J. (1974). *The Limits of Organization*, W. W. Norton & Company, Inc.

[5]Chintagunta, P. K., J. Chu, and J. Cebollada (2012). Quantifying Transaction Costs in Online/Off-Line Grocery Channel Choice. *Marketing Science* 31(1): 96–114. http://www.econis.eu/PPNSET?PPN=688950221.

hierarchical organizations. There are several ways to classify transaction costs, and this is explained for three types in this chapter.[6]

The first type of transaction cost is for matching. It is the cost of searching and identifying who sells what products and services and finding what you should purchase among a vast number of options. When you want to find financial products such as stocks, you can easily find them if you know the name of the company and its prices. However, when you are looking for a designer's skill, you need to know their past track records and work, and their human skills such as sincerity, communication, and proactiveness. In this case, it is difficult to transact in the market. You would be able to achieve a lower transaction cost if you are hiring a designer and order the design in an environment where everyone else knows the designer very well. In other instances in the marketplace, you need to find and decide methods for payment and logistics even after you find a proper product. These selection and decisions around the purchase are also the cost of matching.

The second type of transaction cost arises from uncertainty. If the required tasks and prices are changed after deals are made, then trade in the market will break down due to distrust. In this situation, providers will add a premium on the price as a way to adjust against uncertainty. In the instance of software development, the case where specifications cannot be clearly described beforehand or need to be altered in response to customers' demand comes with high uncertainty. If the cost arising from uncertainty is too high, it is better to assign a member from hierarchical organizations rather than to procure it in the market to reduce transaction costs. Professor Majumdar and Professor Ramaswamy of the University of Michigan examined the effect of uncertainty on downstream integration (integration of firms that are closer to customers) and found that environmental uncertainty, such as with major technological changes, has a positive effect on more possibility of downstream integration. However, behavioral uncertainty, which is related to the third transaction cost, has a three times larger positive effect on the integration.[7]

---

[6]For basic argument on transaction cost, see Willamson, O.E. 1975. *Markets and hierarchies, analysis and antitrust implications: A study in the economics of internal organization.* Free Press, New York.

[7]Majumdar S. K. and V. Ramaswamy (1994). Explaining Downstream Integration. *Managerial and Decision Economics* 15(2): 119–129. doi:10.1002/mde.4090150204.

The third transaction cost concerns incentives. For example, if you order some tasks in the market, your first provider will acquire the necessary knowledge and know-how for your task, and therefore have an advantage in the next and following procurements. In this case, you cannot conduct a fair procurement involving other providers. The transaction cost in this case is a premium that your first provider charges in addition to the usual cost to take advantage of their knowledge and the loss that you cannot purchase a service from a better provider because they do not possess the required knowledge. These transaction costs associated to "opportunism" make it difficult to conduct a task without a certain provider and cause a "hold-up" problem, where you cannot achieve fair purchases in the competitive market.

To prevent this situation, you need to make your providers disclose all information they acquired and report their activity frequently, but this supervision is a serious burden. In this situation, you can avoid the cost if you hire people and ask them to do the task rather than purchasing the service through the market. This is reflected to the situation that bosses can order employees to disclose knowledge and change staffs in charge of the task more easily. Professor Erin Anderson of the University of Pennsylvania revealed, in an empirical analysis of sales staff, that transaction-specific assets and difficulty in evaluating performance by outputs increases opportunism, and integration and goal congruence reduce opportunism.[8]

Whether tasks are conducted through market transactions or in hierarchical organizations is decided by the relative amount of these three transaction costs.

Companies as Mutual Aid Organizations

The transaction cost theory is a strong framework to explain the choice between market or hierarchical organizations, but it misses one aspect, i.e. the theory is based only on the employer's perspective and lacks the viewpoint of workers. It is typically observed in cases where workers have

[8]Anderson, E. (1988). Transaction Costs as Determinants of Opportunism in Integrated and Independent Sales Forces. *Journal of Economic Behavior & Organization* 9(3): 247–264.

incentives to work in corporate organizations, rather than being self-employed or working as freelancers.[9]

One of the reasons to work in an organization is risk avoidance.[10] Even if you have in-demand skills, there is no guarantee that the demand for the skill continues forever. There is a risk that it is impossible to provide the services continuously for various reasons. A study that compared the income of independent entrepreneurs and employees of organizations revealed that some entrepreneurs earn high incomes, but their average incomes are lower than that of employees, while the income of employees converges within a certain range.[11] When individuals trade a business freely in a market, there is a chance to earn high profits, but there is also a great chance that their income will be lower than that of employed workers.

An employment contract mitigates the risk of income fluctuation. In general, employment contracts do not include details of tasks and conditions of rewards, as such that a certain task is worth a certain amount of money. Employment contracts are a system in which workers are paid a stable wage as long as they follow the instructions of their employers. It might be jarring, but employment contracts ensure stable payroll in exchange for the requirement to follow the employer's orders.[12] Workers can quit any time, but their income is guaranteed in exchange with the orders of employers. Essentially, employed workers can feel safe compared to self-employed workers because of their guaranteed stable income.

Aside from stable income, there is a benefit of working in organizations. Hierarchical organizations have an advantage in collective decision-making of a significant number of participants. For example, when business people have to make a prompt decision on a strategy to respond to the environmental change in technology or demand, it takes too long if

---

[9]For details of transaction cost and organizational structure, particularly in the context of blockchain technology, see Takagi, S. (2019). "Does Blockchain "Decentralize" Everything?" In *Blockchain Economics: Implications of Distributed Ledgers*, M. Swan, J. Potts, S. Takagi, F. Witte, and P. Tasca (eds.), World Scientific, Singapore, pp. 25–45.

[10]Milgrom, P. and J. Roberts (1992). *Economics, Organization and Management*, Prentice Hall.

[11]Dillon, E. W. and C. T. Stanton (2017). Self-employment Dynamics and the Returns to Entrepreneurship. *NBER Working Paper Series* 23168.

[12]Milgrom and Roberts (1992), *ibid.*

they have to talk to every person in the organization in a completely equal relationship. On the other hand, they can achieve relatively efficient decision-making if they delegate the rights of decision to those who are good at strategy planning and decision-making (usually it is supposed to be those in the higher levels of an organizations). Therefore, organizations with authority have characteristics in which a specific person makes a decision and others (usually those in the lower level of hierarchy) execute that decision.[13]

In addition, employees in the same organization share common interests and language, and are thus expected to communicate smoothly. Each organization has unique characteristics on what information should be regarded important, whether to report it to bosses, and how to transmit the information. The "communication channel"[14] is a concept in organizational economics that shows a way for employees to collect information and provide it to others in organizations, and this way tends to be sophisticated and made efficient by learning and its repeated use in the organization. In one company, employees can talk to the president in a casual chat at the boss's desk, but in other companies, they have to write a report and receive approval from the group leader, manager, senior manager, and finally communicate with the president. Kenneth Arrow suggested that the learning of the method of information usage is an "irreversible investment" for individuals and "irreversible capital accumulation" for organizations.[15] It takes a significant amount of investment to learn how to handle information in a specific organization by asking their senior employees, bosses, and colleagues. Once you invest in the communication channel, you can lower the cost by using it as much as possible. Organizations can make decisions and take action quickly by sharing the communication channel for collecting and sharing information and by delegating the rights of decision-making.

As seen above, working in a hierarchical organization has the benefits of avoiding individual risks, securing a stable income in economic turbulence, and speedy decision-making involving multiple stakeholders. However, in practice, it is not guaranteed that employed workers can really avoid risks, and that the people in higher levels of organizations are good at decision-making. Considering traditional home appliance

---

[13] Arrow (1974) *ibid.*

[14] *Ibid.*

[15] *Ibid.*, p. 55.

companies such as Toshiba and Sharp are experiencing managerial issues including lay-offs and sell-offs, it is obvious that working in organizations does not guarantee the minimization of risk for workers to zero or secure good decision-making.

Disadvantages of Mature Organizations

Hierarchical organizations have various advantages, but mature organizations sometimes inhibit the climate of innovation. One of the advantages of organizations is smooth and collective decision-making and the sophistication of communication to realize it. Communication channels are an important element in defining how the organization functions, which requires tangible and intangible investment.

The more participants learn and master a method that is unique to the organization, the more cost for communication is reduced and the benefits of working in the organization increase. However, the method unique to the organization tends to coincide with past experiences. If an initial business could happen to be in heavy industry and plant development, its communication channel is not suited for publishing that requires speed, or retail that emphasizes relationship with many customers. A system that is optimized to such initial business models can penetrate and be fixed in an organization, thus making it difficult to respond to environmental changes and to promote new business models.

Kenneth Arrow suggested that rules of an organization are defined by its primary function, but when it does not fit its secondary function, the organization will not function well.[16] In other words, if an organization designs a communication channel that is optimized to the business model from its founding and tries to apply it throughout the organization, including different business models, the development of new businesses in the organization can be intimidating.

This problem of communication channels in organizations exists in the situation where major electronics companies that were successful in manufacturing home appliances could not succeed in digital platforms for consumers. Similar situation can be seen where IT companies whose major field is the customization of software could not expand to packaged software, cloud computing, or web-based services. In the future, similar

[16] *Ibid.*

problems will arise when automobile companies venture into mobility services, and banking companies expand to the FinTech field. The efficiency of communication channels, once the strength of corporate organizations, can turn to a constraint for innovation.

Drastic Reduction in Transaction Costs Through Technological Evolution

The need for transactions across the whole economy has increased drastically since the 20th century. Wallis and North[17] divided the entire economy into the transaction and production sectors, and the transaction sector grew roughly from one-quarter in 1870 to one-half in 1970 in the US. This suggests transaction costs are extracted and externalized from each production industry and grew as independent industries.

The transaction cost associated with trade in the market has reduced drastically in the past decade through technological evolutions. For example, Amazon.com enables users to find the product they want and to trust complete strangers in the transaction. The payment is initiated with "one click" and executed via credit card networks, etc., instead of using postal cash envelopes or sending checks. After the order is placed, the product is quickly delivered to the user's home, even if it is at the opposite side of the earth! These innovations in various aspects such as search, payment, and logistics have drastically reduced transaction costs. Transaction costs can be regarded as a "friction" in transactions, but the reduction of friction enabled us to get a product we want from the best sellers.

Transaction costs on uncertainty, such as ambiguity in business detail, is also being improved. For instance, in software development, when various technologies such as methods of fixing the specifications and development, program language, architecture of modules, project management methodologies, and software to manage products mature, the uncertainty regarding the contract is reduced. In other instances of business process outsourcing, when a pattern of customer relationship management and sales management becomes sophisticated, it becomes easier to outsource

---

[17]Wallis, J. J. and D. North (1986). "Measuring the Transaction Sector in the American Economy, 1870–1970." In *Long-Term Factors in American Economic Growth*, S. L. Engerman and R. E. Gallman (eds.), University of Chicago Press, Chicago, pp. 95–162.

utilizing cloud computing services.[18] It is becoming common to dissolve infrastructural function from services thus establishing standardized services in the form of "IaaS" (Infrastructure as a Service) and "PaaS" (Platform as a Service).

Measures to reduce transaction costs associated with incentives have also been found in the past decade. Visualization of data and review systems have greatly contributed to the reduction of transaction costs on incentives. For example, in the project of outsourcing of call centers, if information on the number of inquiries and answers and users' evaluations are visualized for client companies to grasp the quality of the call-center operations, the information that only contractors know can be reduced. Additionally, in ride-sharing services, the route the driver took can be checked by GPS records, thus validating the action of drivers.

These visualizations of data have the effect to prevent opportunistic behavior. One of the most effective ways to avoid opportunism is the evaluation system, such as e-WOM (Word of Mouth). E-WOM, reviews, and ratings have the power to encourage people to try to act sincerely without the supervision of managers in hierarchical organizations. Various services such as crowdsourcing, i.e. outsourcing of a business to individuals, and ride-sharing, which involves independent drivers, are also examples of preventing opportunistic behavior by the same principle to make the transaction in the market possible.

The sharing economy, in its initial stage, was characterized by an eco-friendly ideology and efficient use of resources, but actually it is the transition from professional services based on organizations to professional services based on individuals. The above-mentioned incentive issues are a problem of "trust" in digitalized market transactions and one of the most important challenges in the digital economy.

Empowerment of Individuals by the Internet

As mentioned earlier, one of the advantages of working in hierarchical organizations is delegating decision-making to those who are good at it, thereby allowing collective decision-making to be efficient. This is based on the time when only people at a higher level in organizations can

[18] Takagi, S. (2017). "Organizational Impact of Blockchain through Decentralized Autonomous Organizations." *The International Journal of Economic Policy Studies* 12(2): 22–41.

collect important information on markets and events and draw a whole picture of the situation. However, since the emergence of the Internet, information now circulates more freely. Compared to the past, anyone who is not at a high level in organizations can collect information on market trends, the latest technologies, and threats from external circumstances. Because "Information wants to be free" — a famous quote from Stewart Brand — anyone can access up-to-date information easily. Young employees in fields, rather than managers in higher levels of organizations, might be more knowledgeable about the trends of technology and services that have a significant impact on management. In other words, information is democratized and accessible to anyone, thus empowering individuals.

It should be noted that there are some aspects of decision-making that should be done by those who are at higher levels in organizations. For example, reducing resources in a section and moving it to other sections cannot be decided or implemented by a lower-level employee in an organization. In other cases, experienced managers would be able to evaluate the latest trend in the context of long-term history. However, the rationality of the notion that "those at the higher levels of organizations are better at decision-making" is relatively weakened. In this context, communication channels that are designed to escalate information to the upper level of organizations and asking for decisions might inhibit the efficient management of the organization.

In addition, the focus of the entire economy has shifted from producing goods to handling information since the late 20th century. Therefore, it is relatively easier for individuals and small entities to conduct important business if they have knowledge, even without massive capital such as factories and industrial equipment. It became possible to start businesses and provide important values for society with a small amount of capital.

Now, it is better for each economic entity to find ways and actions to provide the largest value to society from the viewpoint of efficient resource allocation. Now everyone can optimize the way to utilize their strength as much as possible. Generally, as large organizations tend to prioritize the program based on the idea of "the greatest common divisor," some part of the organization's entities falls into inefficiency, such as when people cannot fully utilize their capability and support is not provided to the people who need it. Behind the scenes, traditional large firms suffer from slumping business, while young entrepreneurs are actively nurturing businesses, particularly in the past decade in

Japan, the reason being those smaller entities can better utilize their potential capacity.

These transforming forces make certain types of business better conducted by small units such as startup firms and individuals creating communication channels that are optimized for the business, rather than through a frame of large organizations. When market mechanisms work, this type of system works much better than the strategy or human resource management with the idea of "the greatest common divisor." Not all organizations are immediately dissolved, but it is important to recognize that information technology is continuously reducing transaction costs and working to optimize resource allocation through market mechanisms.

Dissolution of Organizations by Platform Revolution

The largest force to reduce transaction costs has been brought about by the emergence of digital platforms. Platforms have reduced transaction costs, made information distribution free, and made various resources such as content and people's skills available, thus empowering small entities and individuals. It might seem contradictory, however, that wealth tends to be concentrated on platforms as intermediaries. While it is possible to match buyers and sellers for global transactions, massive-scale platforms like Google, Amazon, Facebook, and Apple receive a huge amount of transaction fees.

Behind the growth of platform firms, the principle of the "network effect" is one of the characteristics of IT services. This suggests that, as the number of users increases, the value of the services also increases. For various services such as Facebook or LINE (an app for messaging that is widely used in Japan), "the more the users the larger the benefit" principle applies. More specifically, the network effect is divided into direct and indirect network effects. A direct network effect is seen in situations where the value of a service increases when the number of its same group increases, as seen in the number of telephone users. Some digital platforms are potentially able to keep their structure as a direct network effect model if they keep users to one group, such as Twitter without advertisers and LINE without third-party stamp providers. However, the dissolution of the economy is driven mainly by digital platforms with indirect network effects, where digital platforms connect multiple, different groups. It generates financial incentives for one of the groups so that the digital platform grows cyclically with boosted investment and mutual effects across user groups. These types of platforms with different groups of entities are called two-sided networks[19] or multi-sided networks.

It might seem contradictory to observe the smallness of economic entities and wealth accumulation through the network effect, but both are caused by the reduction in transaction costs driven by IT. It causes distribution on one hand and wealth accumulation on the other hand.

Simultaneous distribution and accumulation do not mean that the effects are canceled out. On one hand, anyone can easily start businesses as small entities or individuals, and on the other hand, new economic players can benefit from transaction fees by mediating trades between those entities. During the process, the relationship of various players in the economy is reorganized, as shown in Chapter 3. Now is the time that these distributions and accumulation happen simultaneously, and the economy is reorganized.

Dissolving Services and Digital Reconstruction

As discussed earlier, our economy has provided value to the society through the collaboration between colleagues in hierarchical organizations such as companies, but its economic rationality is rapidly changing. As a result of the emergence of new collaborative mechanisms for individuals and small entities, products or services that are specialized and optimized to specific demands would be more valuable to the society.

In addition, small entities and individuals are free from the constraints of other departments, and thus can more easily reconstruct parts of previously analog services as fully digitalized services. In traditional large corporations, entire business processes are designed with analog technologies and are tangled with various rules, sections, and employment contracts; therefore, they tend to face incompatibility with partly digitalized and partly analog services that impede the transformation of business. This is called "Innovator's dilemma," in which innovation is not realized

[19]Parker, G., M. van Alstyne, and S. P. Choudary (2016). *Platform Revolution: How Networked Markets are Transforming the Economy and How to Make Them Work for You*, W. W. Norton.

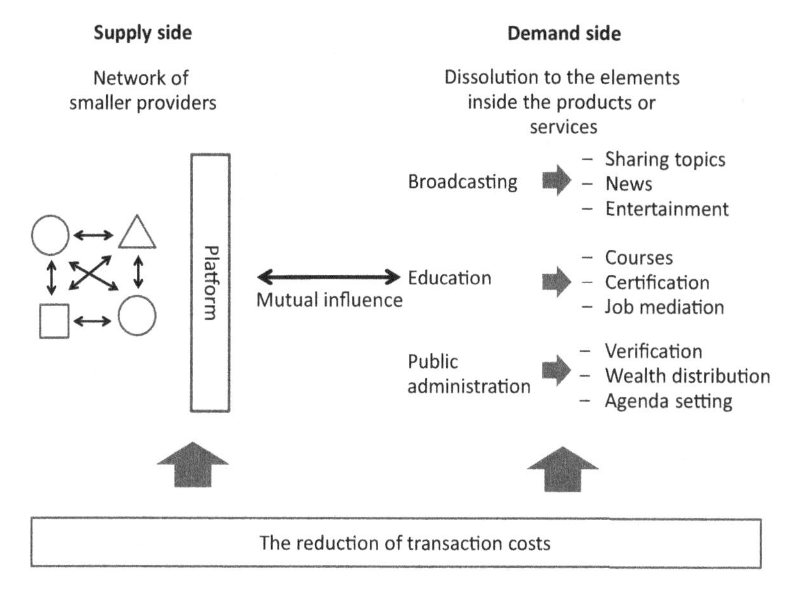

Figure 2-2 Dissolution of supply and demand side

because transforming business models can be a threat to existing businesses.[20]

However, for example, startup firms that do not provide traditional services are not constrained by their past systems and can construct fully digitalized business services. In the financial sector, FinTech firms with small numbers of employees are now providing transformative services that cannot be provided by large banking companies. In this way, as the size of service providers becomes smaller and more individualized, the service itself gets segmented and reconstructed. The concept of this process is shown in Figure 2-2.

For example, television stations used to fulfill viewers' various demands, such as knowing the latest news, and enjoying entertainment such as comedy. However, these functions are now divided, such as sharing common topics via social networking services, knowing the latest news through Internet news services, and consuming entertainment through video-sharing platforms. Thus, each function is dissolved and provided separately.

---

[20]Christensen, C. M. (2016). *The Innovator's Dilemma: When New Technologies Cause Great Firms to Fail*. Harvard Business Review Press, Reprint.

Educational institutions provide various services such as providing classes, verifying and certifying students' abilities, and providing job mediation for students. However, classes are partly available as online educational content, such as MOOCs (Massive Open Online Course). Admission exams partly rely on external examination services such as TOEFL, GRE, and GMAT. Introducing job opportunities has long been provided by various job agent firms.

Many services in public administration can also be dissolved. Public administration includes various services such as certifying addresses and family members, redistributing wealth, and identifying public issues and prioritizing them. If we look at each function separately, the function of certifying addresses and family members could be replaced by blockchain technology. The function of wealth redistribution could also be realized by financial services such as insurance if it is linked with the function of certifying family members and incomes. The linkage between public and financial information would be possible as the movement of linking data across different companies and sectors seamlessly is rapidly evolving, particularly as "Open API (Application Programming Interface)" in the financial sector. In addition, identifying public issues and solving them by prioritization can be better conducted by citizens. It is already partly realized by civic engineers through activities such as the Code for America and the Code for Japan.

The significance of "Civic Tech," which refers to the activities of civic engineers and contributors, proactively acting to solve social issues has been increasing in Japan. Facing the COVID-19 crisis from late 2019, the Code for Japan created the official data visualization website of COVID-19 for the Tokyo metropolitan government.[21,22] This website is published as open-source software, and many other local municipalities have adopted this software to create their data visualization websites for COVID-19. Code for Japan also created a prototype of the contact-tracing application on COVID-19 in a short period of time and published its source code.[23,24] Although its official apps were developed by other firms, it had a great impact on the Japanese government's initiatives to implement those applications. The increasing importance of Civic Tech shows that it benefits flexibility, speed, and dedication to solve social issues.

---

[21] Code for Japan. https://prtimes.jp/main/html/rd/p/000000007.000039198.html.

[22] Tokyo Metropolitan Government. https://stopcovid19.metro.tokyo.lg.jp/.

[23] IT media news. https://www.itmedia.co.jp/news/articles/2005/19/news075.html.

[24] Github. https://github.com/mamori-i-japan.

Dissolving businesses to smaller units makes it possible to rebuild them using digital technologies and also optimize their communication channels. In addition, it is possible to create new business models by combining dissolved elements. Specific ways to dissolve and reintegrate businesses and expand scopes are discussed in Chapter 3.

The New Era of the Personal Industry and the Restoration of the "Invisible Hand of God"

The earlier discussion in this chapter can be reorganized from the perspective of the history of management. Studies on the evolution of corporations have been deepened in the field of management history.

Figure 2-3 shows the evolution of management since 1800.[25] According to Professor Chandler of Harvard University, until the 19th

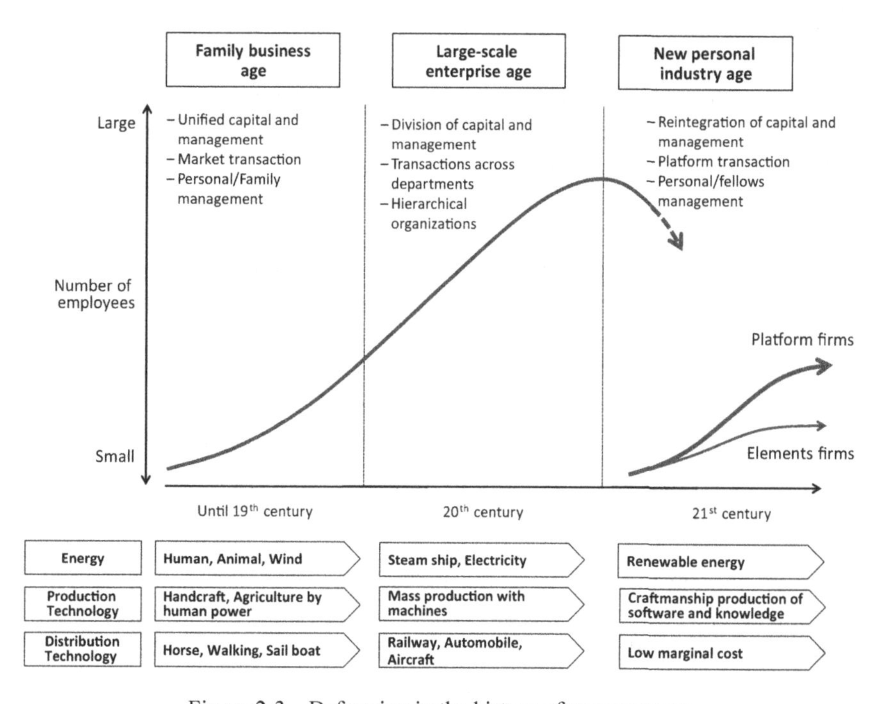

Figure 2-3 Deframing in the history of management

[25] The author composed the figure taking Chandler, A. D. Jr. (1993). *The Visible Hand: The Managerial Revolution in American Business*. Belknap Press (revised), into consideration.

century, management had been conducted through small-scale families and partnerships of individuals. The bottom part of the figure shows the energy, production technology, and distribution technology at that time. Until the 19th century, available energy was natural energy such as wind, animal power, and walking. The available production technology was industry and agriculture using human labor, and distribution took place through horses, walking, and ships. During this time, production and distribution were largely constrained by technological limits; therefore, mass production and sales were not possible. It was the economy that personally managed entities produce, and transactions were conducted within mainly smaller geographical areas.

Professor Chandler also illustrated that, after the Industrial Revolution, drastic change occurred in the US in around 1840. One of the major drivers was the construction of railway networks. Now it was possible to transport products across the country and beyond, so it became possible to efficiently produce and sell massive amounts of products. Energy was transformed through steam power and electricity. The technological revolution enabled mass production with machines and mass transportation of products with trains, automobiles, and ships to distant destinations.

In these circumstances, it is difficult for individuals and family companies to take advantage of such technological innovations. Production equipment requires large-scale capital, the funds for which are provided by investors and financial institutions. On the other hand, as organizations become larger companies with multiple departments, these companies are managed by hired professional managers, not owners, separating capital and management. Professor Chandler described the coordination of transactions across multiple departments by managers as a "visible hand." Large corporations transformed transactions in the market through the "invisible hand," as described by Adam Smith, to coordination by managers. Thus, the age of large corporations has emerged and continued until recently. These are the outlines of the management history drawn by Professor Chandler.

Management in Japan until the 19th century was also mainly personally managed, such as with merchants or agriculture, and large-scale firms were rather the exception. These exceptions can be seen in the cases of Mitsui, a *Gofuku* (traditional Japanese clothing) wholesaler; Sumitomo, a copper mining firm; and Konoike, a money exchanger, through *the Edo* period (1603–1868). Even in the Meiji period (1868–1912), 70% of the population was farmers, and manufacturing took the form of household industry. Since the late 19th century, the Japanese economy also started to introduce large machinery for textile spinning, forming the modern shape of corporations.[26] This structure of the industry led Japan to become one of the most successful economies in the world in the late 20th century.

However, this trend has been changing since 2000. First, important value in the economy has shifted from physical goods such as industrial products to software and information. These intangible assets are produced based on sophisticated knowledge embedded in individuals with a style similar to that of traditional artisans. Second, the distribution of such software and information is conducted at a significantly lower cost using Internet infrastructure, such as putting the information on websites or sending it enclosed in email. As Jeremy Rifkin described in his book *A Zero Marginal Cost Society*,[27] with the reduction of marginal cost, any entities without large-scale capital can infinitely distribute value to the world. As a result, firms with relatively smaller number of staffs can now provide values disproportional to their size. It can be considered a

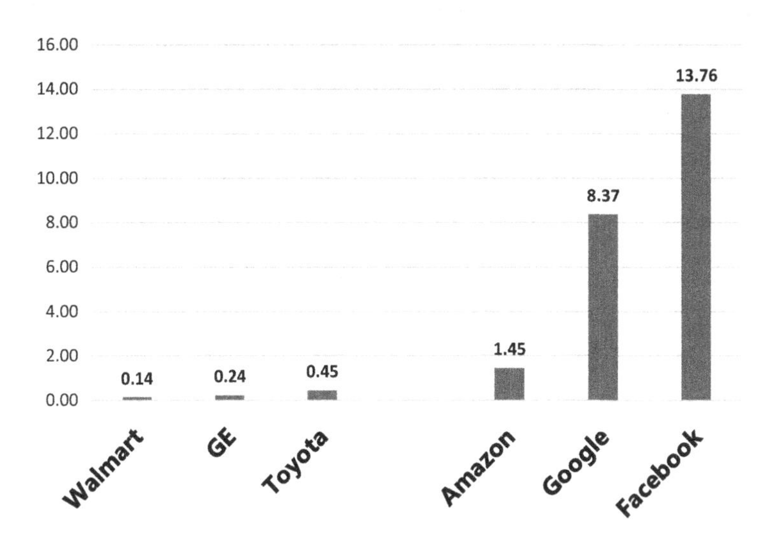

Figure 2-4 Firm value per employee, million USD

[26]Miyamoto, M., K. Okabe, and K. Hirano (eds.) (2014). *The 1st Step of Business History*, Sekigakusha (original in Japanese).

[27]Rifkin, J. (2014). *The Zero Marginal Cost Society*, St. Martin's Griffin.

renaissance of the "invisible hand of God." Entrepreneurship in the modern age does not require much capital; therefore, capital and management are reintegrated as individuals provide capital as well as the management skills to conduct businesses. We have entered the "new personal industry age" following the ages of the family business and the large corporation.

In the future, there will be two major aspects of industrial organizations. One is the economy with smaller business units that provide services. The other is platform-based firms that mediate smaller business units. Platform firms tend to become larger because they mediate trust in transactions across individuals and acquire massive amounts of data that generate important insights (see Chapter 6). Even in this case, the number of employees will not become as large as that of traditional manufacturing companies. This is because firms with a small number of employees can provide services to a massive number of customers in the digital economy. Figure 2-4 shows that platform firms can generate larger firm value with significantly smaller number of staffs.[28]

As seen in this chapter, the principle of deframing suggests the end of the age of gigantic, hierarchical organizations that has continued since the 19th century. The economy is transforming into a new form where smaller entities provide world-class services based on their knowledge and low marginal costs, collaborating with other entities. It does not necessarily mean the end of large corporations, but flexibility and optimization for each business domains are becoming far more important even for them. Deframing suggests a significant turning point for management and organizations.

[28] *Source*: Yahoo! Finance. As of November 12, 2018.

Part II

The Process of Deframing

Chapter 3

Dissolution and Reintegration

Transformative Tech Services and Deframing

In this chapter, we will dive into the contents of deframing using specific examples.

The first element of deframing is "dissolution and reintegration." This means to specialize in an element in the existing frames and recombine, reassemble, and reintegrate these elements to create new values. Dissolution is conducted from an objective perspective, and new elements should be combined to create services realizing the economy of scope.

For example, educational institutions such as universities provide a variety of functions such as classes, clubs, certifying academic abilities, job recruitment, and a sense of community. Banking firms provide deposits, remittance services, loans, investment opportunities, and the provision of credit information. The telecommunication sector provides various functions such as telephone, data communication, and messaging (Figure 3-1).

One of the countries where dissolution and reintegration is the most advanced is China. In China, specifically in cities such as Shanghai, Hangzhou, and Shenzhen, various IT services such as QR-code–based mobile payments, ride-hailing apps, and Online-to-Offline (O2O) commerce services are proliferating on a massive scale and at a rapid pace. People can experience the advantages of digital technology and innovation in various aspects of life, such as in retail, mobility, and dining. However, when we look at those services closely, we can understand that they follow the principle of deframing.

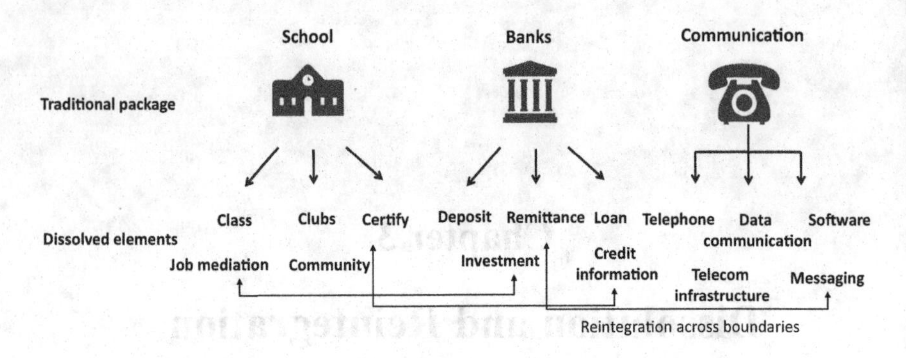

Figure 3-1 Dissolution and reintegration across sectors

One of the examples is the Chinese tech giant, Tencent, whose market cap is the sixth largest in the world.[1] It was founded in Shenzhen, a southern Chinese city, with its first service being QQ, a type of Internet-based chat system. QQ has evolved into "QQ game," a social game, while Tencent also launched a blog service. Taking advantage of its strength in social networking systems, they launched WeChat, a messaging application. WeChat itself is a similar application as Facebook messenger and LINE, which supports communications working on smartphones.

However, their services do not end in communications. They expanded the services to WeChat Pay, a financial payment service on smartphones. On WeChat, users can send money at almost no cost, such as a situation where friends send money to each other and split bills. In China, people can pay using WeChat in most stores, taxis, or vending machines. When a service has a massive number of users for its social networks and payment infrastructure, many other firms will want to conduct businesses on that platform. Tencent now provides "mini-programs," which is a third-party application platform that functions in a WeChat application. In other words, Tencent adopts a multi-layer strategy that consists of the Apple or Google platform at the bottom, Tencent applications in the middle, and mini-programs on top of it. Tencent is now providing broader overarching services, including e-commerce, ride-sharing, and financial services. Tencent deconstructed communication services

[1]2020 Fourth quarter. *Source*: Wikipedia.

and specialized on "messaging in social networks," expanding its economy of scope afterward, adding the elements of online remittances, cloud service, and e-commerce.

Alibaba, another Chinese Internet giant, exhibits a case of expanding the economy of scope starting from dissolution and specialization. Alibaba is now the eighth largest company in the world,[2] but it was originally just a business-to-business (B2B) online commerce platform, Alibaba.com. This platform is used by merchants to buy goods from producers and wholesalers. Alibaba started its business specializing in B2B platforms and expanded with a consumer-to-consumer (C2C) platform, Taobao. It is a service that enables commerce between individual sellers and buyers, as seen in Mercari. Alibaba further founded Tmall, a business-to-consumer platform, where consumers can buy goods from merchants, similar to Amazon.com and Rakuten. As seen here, Alibaba started business from a specialized B2B service and expanded to C2C and B2C businesses.

On the other hand, a payment mechanism is required to make online commerce platforms function. "Alipay" was the payment function for Alibaba's platforms, and the popularity of this service supported the overall growth of the platform. It is a mechanism that supports smooth trade between unknown participants without established mutual trust. In deals made in the online market, after a buyer makes a payment for the goods, there is a risk that a seller does not ship goods to the buyer. On the other hand, there is also a risk that a buyer does not make payment if a seller ships it first. The escrow service of Alipay solved this problem of the lack of trust between stakeholders by keeping the payment as a deposit until the buyer is satisfied with the products.[3] Thus, Alibaba succeeded in substituting the trust of unknown traders and creating a platform to grow.

This payment system, Alipay, had an unexpected byproduct. Alibaba was able to add many functions utilizing its massive amount of payments and deposits. For example, they could use the system for payment between individual users as well as between physical retailers and buyers. In the end, Alipay proliferated as a payment system that is used for many different purposes. Alipay and WeChat Pay became popular in and also

---

[2]2020 Fourth quarter. *Source*: Wikipedia.

[3]Botsman, R. (2018). *Who Can You Trust?* Public Affairs.

partly outside of China, and consumers can be observed using these services for manifold transaction in any Chinese city.

Transformation from Payments to Mediating Trust

Alipay's evolution does not stop there. If the service can manage payments, it can utilize the information on who is conducting what business with whom. The payment data can reveal if users are paying bills for water and electricity properly, what they are paying for, to whom they are making payments, etc. In other words, the service can estimate each user's financial capacity, repayment ability, and in more abstract terms, the level of credit. A service for payment systems now turns into the infrastructure for credit information. Thus, "Sesame credit" was born.

Now, for users of Alipay, a "Sesame credit" score is automatically calculated based on the profiles and usage of Alipay. Its score, if high enough, enables users to take advantage of their credit levels, such as free deposits at hotels and complimentary services. In fact, a hotel in Shanghai that I stayed at during the course of my research was also offering free deposits based on the Sesame credit score. Currently, the businesses of Alipay and Sesame credit are conducted by Ant Financial, a company in the Alibaba group. Alipay was born as a payment system for Alibaba.com, but later emerged as an independent payment service and further reconstructed itself as a credit infrastructure and entered the next cycle of expanding the scope of the economy. Alibaba also designed incentives so that the more users use Alipay, the higher the score of Sesame credit they get.

Currently, various companies are providing credit scoring services, such as Jingdong's Xiao bai credit. Xiao bai credit is also used in situations such as renting toys without deposits to users with higher scores. Thus, rapidly growing Chinese IT companies have transformed business models through "dissolution and reintegration." These credit score services have been introduced also in Japan, such as J.Score and Yahoo! Score. However, these credit scores also raised the concern on the privacy, and some of the services had no choice but to be terminated. Privacy issues related to credit score is discussed in Chapter 6 in more detail.

Restructuring of the Industrial Structure by Deframing

Dissolution, one of the key elements of deframing, has a long history, particularly in the IT industry. Professor David B. Yoffie of Harvard Business School and Professor Michael A. Cusumano at the Massachusetts Institute of Technology described how the industrial structures of computer hardware have changed from vertical to horizontal in their book, *Strategy Rules*.[4] According to them, computer manufacturers such as IBM and DEC were developing everything from CPUs, operating systems (OSs), and software, vertically integrating from development to distribution and sales. Facing this situation, Bill Gates, noticing the importance of software, came up with a model to provide the same software to all manufacturers by approaching them from a horizontal perspective. Since then, the computer industry has transformed into a horizontal industrial structure in which each firm provides horizontal layers such as CPU, OS, business software, hardware, distribution, and sales.

In other words, firms specialize in what they are best at, such as CPU or OS. Vertical integration can ensure interoperability across products, but at that time modularization, a standardization of specification on the interfaces of components inside of computers, was progressing, and the industrial structure has rapidly shifted from vertical to horizontal.

Since the 1990s, not only in hardware manufacturing, but also in software development, business processes have been disintegrated through a form of "offshore outsourcing." For example, firms outsourced parts of software development processes to countries with lower labor costs, such as China, India, and Vietnam. Business process outsourcing (BPO) was also actively used for labor-intensive processes such as call-center operation, data entry, and accounting.

Since then, outsourcing has continuously evolved by freely cutting parts of businesses and outsourcing them through cloud computing for accounting, sales management, and IT infrastructure. My book *Reweaving the Economy: How IT Affects the Borders of Country and Organization* (2017, University of Tokyo Press) analyzed the impact of such outsourcing and cloud computing on the Japanese economy.

---

[4]Yoffie, D. B. and M. A. Cusumano (2015). *Strategy Rules: Five Timeless Lessons from Bill Gates, Andy Grove, and Steve Jobs*, Harper Business.

Since the late 2000s, as a form of crowdsourcing, new types of purchases of various businesses such as design, data entry, translation, and software development from individuals have emerged. Its basic mechanisms are not different from outsourcing, but their economic players have become smaller from companies to individuals who take orders of jobs as economic entities.

However, "outsourcing" is no more than the relationship of sales-ordering, which is considered as an extension of traditional hierarchical organizations in the sense that purchasers have the absolute right to order detailed specifications and make decisions. The business process was dissolved within the existing business scopes, and industrial sectors and firms' positions were not affected by outsourcing.

Rebuilding Services with "Digital-First"

In traditional outsourcing, dissolution and reintegration were conducted within the existing distribution of tasks in firms. However, digital technologies are enabling the dissolution and reintegration of business processes in far more detailed units and across industrial sectors.

Based on the deframing principle, it is possible to extract individual elements from existing frames and rebuild as a "digital-first" service. Simply put, digital-first services are services whose components are fully enabled and integrated with digital technologies. For example, it takes a form of a service that completes the whole process with digital technologies such as user authentication, payment, record management, receipt management, delivery management, and post-purchase services. Certainly, some elements such as delivery, transportation, and manufacturing retain physical elements, but information management for those processes can also be done with digital technology. In the deframing standard, it is necessary to enable fully digital services, avoiding the requirement for users to send envelopes or to fill in paper documents.

Another element of digital-first is to link with personal identity. Rather than providing services to the unanimous mass market, it is a service knowing who is the customer. There are various ways to link with customers' identities, such as installing smartphone apps, SMS authentication, e-mail address registration, name registration, credit cards, and banking account registration. Depending on the method, the degree of linking may vary.

As examples of digital-first, suppose you are opening a bank account. If a required document is provided with an electronic file such as a PDF, but you must print and submit it, there is little benefit for you. If certification information in the public sector is directly transmitted to financial institutions that need it, there is no need for paper documents. If it is possible to process it using only a simple operation for smartphones, the user benefit is enormous. In fact, there is a possibility that this transmission of information itself can become an important business element. In addition, this function of transmitting information with users' consent can be expanded to provide real estate agents, corporate registries, and job application processes the benefits of the economy of scope. Conversely, it is also possible to transmit credit information in financial institutions with users' consent to aviation companies and hotels, so that users can enjoy discounts based on their credit information. Small elements such as transmitting information with users' consent represent great business opportunities.

The reason for not digitalizing entire traditional services but focusing on small elements of business is that traditional frames of packaged services are defined by the conditions of the analog age, and it is not necessarily an optimal combination. In addition, traditional services are designed "analog first," and analog elements are intertwined with each other, thus making it difficult to rebuild the entire service digitally. Therefore, completely different players have the advantage of rebuilding a digital-first service of a limited element and provide it to fit the expectation of modern users on user experiences, thus quickly delivering transformative services for customers.

"Z-shape" Strategy of Deframing

Then, should deframed services be limited to a fragmented, small scope of business? It does not necessarily do so. It is possible to rebuild scalable services based on a new way of thinking. One of the essential concepts to think about business strategy in the deframing age is "the economy of scope."

Generally, the economy of scale refers to the idea of reducing the average production cost by producing and providing uniform goods in high quantities. In contrast, the economy of scope refers to the idea that when firms produce a variety of goods or services, there is a part that can

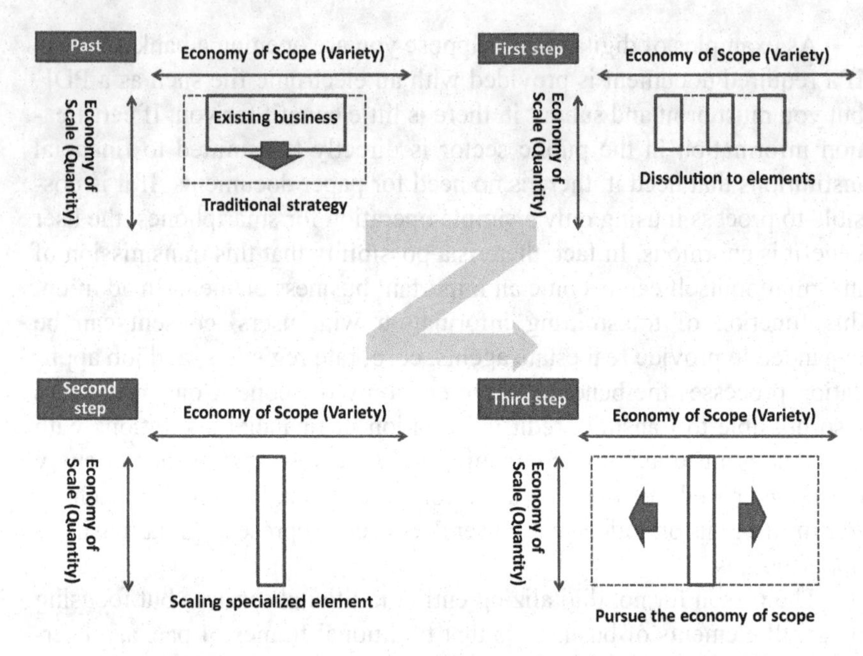

Figure 3-2 "Z-shape" steps of dissolution and reintegration

be communalized; therefore, they can reduce the average cost of the overall product line.

In deframing, when firms start from dissolving elements and expand their business, the economy of scope plays an important role. Figure 3-2 illustrates the "Z-shape" steps that show the flow from dissolution to the economy of scope. The traditional business model (upper-left) shows the basic growth strategy where firms try to expand the scope and scale gradually simultaneously. In the deframing age, firms' growth strategy takes three steps.

The first step (upper-right) dissolves existing businesses and services to small pieces. If it is a retail store, it can dissolve elements to selecting goods, displays, payments, advertisements, logistics, etc. Logistics can dissolve various elements such as physical transportation, logistics adjustment, order management, insurance, and international trade management. The first step is to dissolve the elements in this manner.

In the second step (bottom-left), among the dissolved elements, firms choose the one in which they are the most competitive, and that has the greatest potential to be improved by digital technology, and rebuild the element using the digital-first principle, and specializes the element.

In exchange with the narrow scope of business, the firm should pursue market share and the number of users on the service. For example, firms should achieve a major or leading share in the mobile payment market. During this stage, the important thing is to acquire enough users for the expansion in the next step. Sometimes, it takes the form of being free of charge for end-users while combining the strategy of a two-sided market or simply using the resource from funding from investors.

In the third step (bottom-right), firms expand their scope by leveraging the specialized element. For example, firms can expand from mobile payments to e-commerce, social networks, and mobility. In other cases, it is possible to start certifying academic records and expand to job matching and educational content services. An important point here is that it is possible to combine with businesses in a totally different field from what they started from. Tencent, as seen before, started with the online chat system in the communication sector and social games in the gaming sector and added payment services, thereby merging with the financial sector. In addition, it expanded to application platforms by "mini programs," which are third-party apps within WeChat. Alibaba similarly evolved from the B2B market to the C2C market, adding Alipay in the financial sector, and Sesame credit as a credit information service. If it is reconstructed digitally first, it is possible to combine elements in other sectors, crossing the border of fields.

It is important to acquire a number of users by a small piece of an element that is rebuilt by digital-first ideas and expand from this element. Therefore, when firms reach the third step of deframing, their business is reconstructed as digital-first services in a completely different sector from where it was operating in simple expansion (upper-left).

Social Commerce as Dissolution and Reintegration

The development of the business model and transformation of the industrial structure based on "dissolution and reintegration" can be observed in various cases other than Tencent and Alibaba. As an example, RED, a word-of-mouth app based in China, was founded in 2013 and has provided a service to share electronic word-of-mouth (E-WOM) on smartphone applications. It had more than 70 million users in early 2018.[5]

---

[5]SankeiBiz. https://www.sankeibiz.jp/business/news/180119/prl1801191802119-n1.htm.

RED has a function to share how to use a product, reviews, and daily blogs as well as serving as a social networking service, enabling users to follow other favorite users. However, responding to the growing demand to buy the goods reviewed in the app, the function of selling the goods was added in 2014.[6] Thus, RED has developed a new application field, "social commerce."

A similar function has also been employed in Instagram since 2017. Instagram was originally an app to share locations and pictures, but recently it has been used by stores posting products, and influencers such as celebrities and influential users promoting particular goods. It has evolved from just sharing pictures and making comments to each other to a platform for promotion and advertisement of products. Instagram further promoted this trend and started "Instagram shopping," a service that enables the purchase of goods shown in posted pictures. It shows the expansion of scopes from picture sharing to e-commerce.

RED expanded from E-WOM to e-commerce, and Instagram expanded from photo sharing to advertisement and then e-commerce. Even if the starting point is a small scope of business as a niche market, if they acquire enough users, they are ready to expand to benefit from the economy of scope.

Deframing in Logistics and E-Commerce

The development of IT innovation in the Chinese market is worth paying attention to. One of these interesting services is Meituan (formerly Meituan Dianping), an e-commerce platform whose strength is delivery services. While Tencent and Alibaba have demonstrated significant presence in smartphones, Meituan has a significant physical presence in cities. In major cities such as Shanghai, Hangzhou, and Shenzhen, you can see workers riding electronic bikes wearing Meituan jackets and transporting packages.

Meituan was originally a service started from E-WOM and then evolved to a website to sell coupons for stores. After that, it evolved to a service where transporters go to a store to pick up food that customers order from the smartphone application. The food is delivered in around 30 minutes to 1 hour, and you do not have to leave your house to buy

---

[6] @Press. https://www.atpress.ne.jp/news/147785.

goods or foods. Meituan is evolving to a general e-commerce platform. The platform deals with not only food delivery, but also provides a wide range of services such as delivering goods from supermarkets and convenience stores, movie tickets, hair salons, hotels, trips, karaoke, and bars. It also mediates various services such as housekeeping, moving, fitness clubs, car repair, tutoring schools, interior construction, wedding dress, and medical services in the platform.

A remarkable part of Meituan is that it focused on the element of transportation and pursued the economy of scope based on it. Now is the time that everything can be processed digitally, but transportation remains a bottleneck for convenient services. Solving transportation issues using a large number of human resources is unique to the Chinese economy, which has a massive population. It is an interesting case of leveraging the transportation capability to be competitive in e-commerce, because transportation is usually considered as a downstream of business flows and is labor-intensive. However, given the high entry barrier in the logistics segment, it can be a key for success in the digital platform business. The importance of logistics in digital platform business is becoming apparent, as seen in the announcement of alliance between Z Holdings, that holds several e-commerce services such as Yahoo Japan and ZOZO, and Yamato Holdings, a major logistics company in Japan, on March 24, 2020.[7]

Meituan is also interesting from the viewpoint of the platform economy. Most platform businesses mediate transactions across two different groups of entities, such as stores and customers, music creators and listeners, application developers and users, and drivers and riders. They are typically called a "two-sided network" because they connect two different groups.[8] Major IT firms such as Google, Amazon, and Apple used this business model to achieve rapid growth, and it has been the source of competitiveness (Figure 3-3).

On the other hand, Meituan constructed a "three-sided network" that consists of three parties of stores, customers, and transporters (Figure 3-4). While they do work for Meituan, it is up to them whether they take delivery orders. According to a transporter, the attractive point of this job

---

[7]NIKKEI. https://www.nikkei.com/article/DGXLRSP531534_U0A320C2000000/.

[8]Parker, G. G., M. W. Van Alstyne, and S. P. Choudary (2016). *Platform Revolution: How Networked Markets Are Transforming the Economy and How to Make Them Work for You*, W. W. Norton.

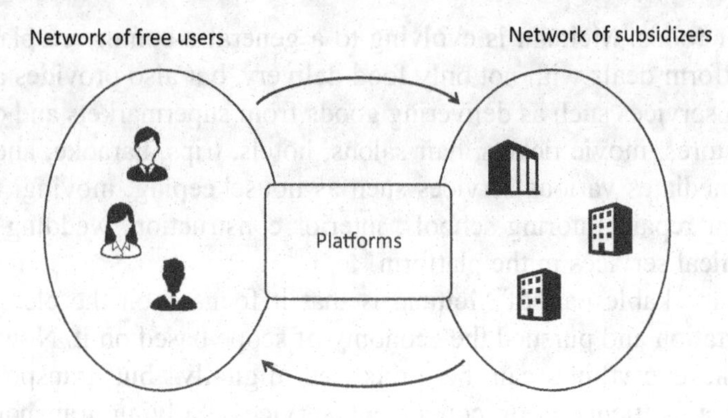

Figure 3-3 Network effect of platforms
Source: Author, referring Parker *et al.* (2016).

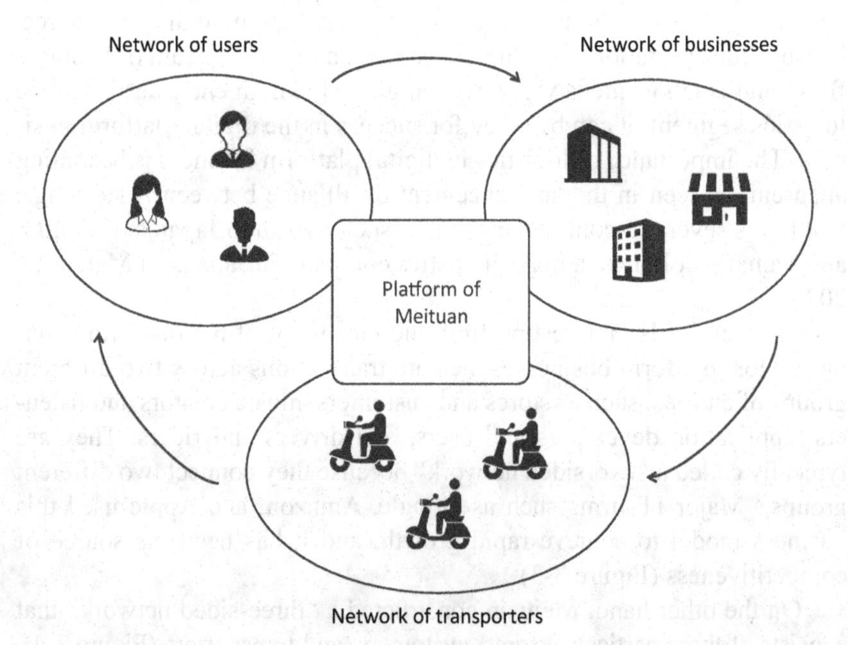

Figure 3-4 Network effect of three-sided network

is the freedom on the tasks. This suggests that their work is discretionary, like for drivers of Uber and DiDi, and they work as self-employed or as freelancers who trade on platforms freely than employees under command and control. This is related to the third element of deframing: individualization.

What kind of significance does the three-sided network have? Until now, Meituan has constructed a large-scale network and is competitive in the market, and latecomers that have to construct the network in the future have a disadvantage in this regard. Considering the difficulty and complexity of the network, the network externality might be larger in a three-sided network than in traditional two-sided platforms. On the other hand, the Chinese labor market has high liquidity; therefore, there is a risk that transporters would shift to other job opportunities that have better conditions. Additionally, because of delegating the key factor of transportation to transporters, there is a risk that the platform cannot secure sufficient logistics capacity. This may cause a situation in which customers want to order, but transaction is not made because there are no transporters. If others construct similar platforms in the US, Europe and Japan, the challenge is whether they can secure enough transporters. Meituan's business model that enables a logistics network using platform technology might be unique in China, which has a large population. In countries with smaller or decreasing populations, drones and autonomous cars could be utilized with platform mechanisms.

Impact of Deframing on Existing Business Models

Dissolution and reintegration affect existing business models. The transformation seen in Meituan's case involves Online-to-Offline (O2O) transformations, which affect the role of physical retail stores as a result of reintegrating the value chains of commerce. In the past, retail stores, such as convenience stores and supermarkets, were places where people visit to buy goods. However, because of the growing use of delivery services, far fewer people visit physical stores, and the role of stores is changing to be more similar to logistics facilities.

Suning Xiaodian, where I visited with Takeshi Yamaya, an expert writer on Chinese IT innovation, is a convenience store in China that is experiencing such a transformation. Suning Xiaodian is a convenience store that is located within a home appliance store chain, Suning.com, that previously acquired the Japanese competitor Laox. It operates as a logistics center as well as a retail store for foods and daily necessities. The store itself is located in a convenient place facing a street, but its store was used as a place from where goods are delivered based on the order from the Internet. Its interior is also like a logistics center rather than a retail store for customers.

Alternatively, Hema Fresh, an advanced supermarket funded by Alibaba, focuses on delivery services and transport of goods within 30 minutes to a radius of 3 km. For this purpose, equipment similar to a monorail is placed around the ceiling, and when order is placed via the Internet, goods and foods are carried through the monorail to the logistics center in the store. Transporters with bikes are waiting at the logistics center, who carry the products to destinations such as homes and offices. It is a relatively large-scale supermarket, but the structure of the stores is being transformed by the development of transportation services. Besides Hema Fresh, many other competitors such as the Ella Supermarket of Meituan and Le Marche, operated jointly by Carrefour and Tencent, provide cutting-edge services such as facial recognition payment.[9]

With the development of delivery services, there is a possibility that consumers will not go to the physical stores. The author met young people in China who said that they do not go to supermarkets. To respond to this situation, some supermarkets in China provide interesting systems to attract customers. For example, if you stand at a big screen in a store, the machine takes a picture and your picture is put into a cartoon that is shown on the screen, and you can also download the cartoon picture in your smartphone. You can also see your picture on the screen virtually wearing clothes that are sold in the store. These systems reflect the fact that stores are no longer just places to buy goods but also function as places of entertainment where customers just see and check goods, but buy the items online.

These are some examples that show the transformation of social functions of each entity as a result of "dissolution and reintegration." In the age when anything can be bought by ordering on the Internet and having it delivered, the role of retail stores is changing to that of an entertainment and logistics center. The origin of the change in the retail sector in China is "New Retail Strategy," initiated by Jack Ma, the founder of Alibaba, in 2016. Through measures such as the merging of online and offline (physical store), automated convenience stores that utilize technology transformed the experience of retail for consumers and sellers, thereby impacting the industrial structure.

As mentioned earlier, innovations arising from the Chinese market show that innovators start from dissolved functions such as payment,

[9]RetailDetail. https://www.retaildetail.eu/en/news/food/carrefour-and-tencent-launch-high-tech-store-shanghai.

Social Networking Service (SNS), and e-commerce, and then pursue the economy of scope by combining with logistics, commerce, and credit information, thus building new social platforms. The Chinese market is characterized by the general interest on business and entrepreneurship, and relatively flexible and liquid labor market, and these characteristics are comparable to that of the US. Its business-creating process will expand to other countries sooner or later, as long as they share similar economic characters.

Seamless Service Realized by the Application Programming Interface Economy

In "dissolution and reintegration," we adopt a broad perspective such as business domains and companies. However, by observing it more closely, dissolution and reintegration in micro-perspective can also be deemed crucial, particularly in using application programming interface (API)

The API is similar to a "doorway," allowing digital applications to communicate with each other: it allows different systems to exchange information, thus making it possible to coordinate business-related tasks. These APIs have a network that extends in all directions, and the API economy is one wherein information and value are distributed. In Japan, due to national policies,[10] financial institutions (particularly banks) are promoting "open API" initiatives to make APIs publicly available to ensure that applications developed by third parties can use the bank's information. Therefore, household account or investment applications would be able to access the account balance of users that are managed by banks through appropriate security controls and provide increasingly seamless services. This process would be increasingly seamless because users would not have to manually download data from individual bank's website and upload it into the household accounting software. Moreover, this ensures that data can be linked from the beginning to ensure the user can receive integrated services.

---

[10]The Financial System Council's "Working Group Report on the Enhancement of Payment Services — Strategic Initiatives for the Enhancement of Payment Services" (Published on December 22, 2015, original in Japanese) and the government's "Japan Revitalization Strategy 2016 — Towards the 4th Industrial Revolution" (Cabinet decision made on June 2, 2016, original in Japanese).

Although banks may be able to offer various services such as investment, financial records, and tax filing, due to the rapidly growing use of smartphones, the development of significantly convenient new services is progressing at an enhanced speed on a global scale. From a cultural perspective, setting out to develop flexible applications is a difficult challenge for traditional banks, which have specialized in the accurate and safe management of deposit balances and interest rates. Therefore, the value of the bank can be increased by linking with applications developed by flexible and effective third parties, such as startups.

As users become accustomed to new and convenient experiences by digital technologies, users will demand an equivalent interface for all services, such as government services in the near future. From the user's perspective, they should be able to easily download their certificate of residence, instead of visiting a government office, thus entirely eliminating the paper-based system. To achieve this, the government's system must link to applications made by various third parties and a mechanism must be set in place for safely sending resident certificate data. Moreover, for services that require a resident certificate (e.g. when opening an account at financial institutions), a mechanism must be set in place to send the data directly from the government office to financial institutions (Figure 3-5).

Services that require users to submit a resident certificate do so to confirm that the address or name declared by the user is correct and do not necessarily require the paper version of the resident certificate. If it becomes possible to obtain results by using dynamic queries for systems that retain the information, then based on the purpose, people would not have to retrieve their residence certificate. Currently, the API economy is

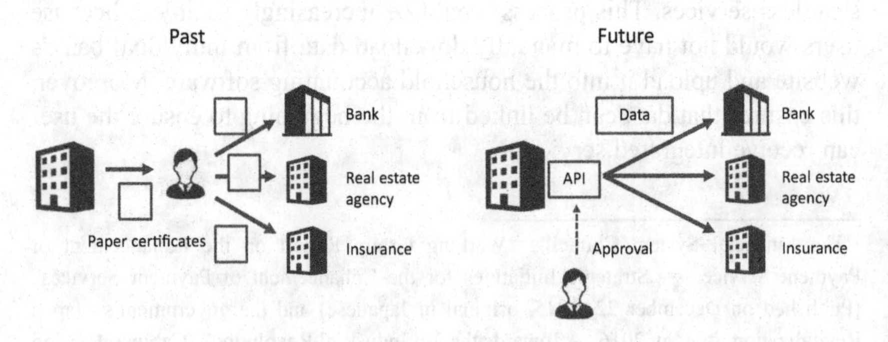

Figure 3-5 Seamless network with APIs

an indispensable element for realizing services that we may take for granted. Dissolution can make traditional businesses to become minimized functions of APIs, and services would be replaced by a network of those APIs in the extreme end.

Deframing in Energy Sector

Due to APIs, deframing in the electric power field is also related to dissolution and reintegration. Until now, electric power has been generated at power plants by using large-scale power sources, such as thermal, hydroelectric, and nuclear power, and a mechanism has been set in place to transmit and distribute this energy simultaneously. This centralized system is because it was more efficient to use electricity derived from fossil fuels and nuclear power, which is difficult to handle. However, amid these technological changes, such as a global shift toward renewable energy (RE) that can be used indefinitely (e.g., solar and wind power), "centralized" power transmission systems are no longer the most efficient systems.

RE can be easily generated using solar panels at home, and even small businesses can easily enter the power generation business by creating energy farms. However, the power sources will be geographically dispersed. It would be extremely inefficient to extract the electricity generated in a scattered area, collect it at a central location, and then redistribute it. Therefore, it would be advantageous to create an exchange for electricity in a small area, such as a local area or neighborhood.

The implementation of this type of local/neighborhood power exchange is rapidly progressing, due to it being combined with virtual currency and blockchain technology. A technology called the "digital grid," which was developed by a research team at The University of Tokyo, can identify where and how electricity is generated. It will become possible to purchase electricity only derived from solar power generated in the same neighborhood. Furthermore, by using blockchain technology to prevent data tampering, the technology is being developed to accurately record the amount of electricity generated and consumed, and to settle payments for the amount of electricity transacted.

As seen above, it is now technically possible to carefully select the type of power purchased, such as solar, wind, nuclear, and petroleum power, compared with the aspects that have been previously offered within the substantial framework simply called "electric power." By examining these technological changes in power generation methods and the hidden needs of consumers seeking sustainable living and business options for the global environment, it become possible to provide individually optimized energy. In addition, individuals and small businesses will also be able to participate in the generation and provision of energy.

A major feature of conventionally generated electric power is the principle that a stable equilibrium exists between demand and supply simultaneously, and that there is no discrimination because the source cannot be identified. Therefore, it was necessary to control the huge electric power system through central management. However, high-performance batteries that can store electricity such as used in electric vehicles (EVs), are rapidly evolving, and the idea of a simultaneous equilibrium between demand and supply is no longer a given.

Charging time has been an issue for EVs, the use of which has recently been rapidly increasing globally. Fully charging an empty battery can take 30 min or more at the fastest. This may not be an issue if the driver has errands, such as shopping, to attend to, but compared with the experience of refueling vehicles using gasoline in approximately 5 min, the additional time required for EVs can be stressful for users in the middle of a drive or in a rush. In this context, EVs with replaceable batteries are taking attention. For example, one of these systems is provided by a venture company called "SKIO" in China, and similar to replacing the battery in a mobile phone or watch, the battery replacement can be completed within a few minutes by simply switching out the battery. In Chinese cities, it is also possible to find automated sharing units for charged batteries; this involves temporarily borrowing a pre-charged battery and using it to charge one's smartphone.

This indicates a partial break from the requirement of a stable equilibrium between demand and supply of electricity, and transforms electricity into a portable commodity. By combining the identification of electric power using the digital grid, resources can be disassembled into the businesses involved in power generation and storage, thereby allowing various stakeholders to become involved. If the costs of electricity storage and logistics are further improved through technological progress, transporting batteries charged using solar panels in the desert, for use in many other economies, will become a possibility. Perhaps it is time to radically rethink the cost of electrical power production and supply, as well as the valuation of energy.

Technological Frontier is the Collaboration of Autonomous Subsystems

In dissolution and reintegration, the method through which disassembled subsystems are connected and how they cooperate is crucial. This type of cooperation between systems has been promoted as Industry 4.0 in Germany, with a focus on improving productivity in the manufacturing industry. This is presented as "Society 5.0" in Japan, as described in the government's Basic Plan for Science and Technology. Society 5.0 is a human-focused society that balances economic development and focuses on solving societal issues by fusing cyberspace and physical space. This idea is called Society 5.0 because in human history, the hunter-gatherer society represents Society 1.0, the agricultural society represents Society 2.0, the industrial society represents Society 3.0, and the information society represents Society 4.0. If the beginning of the information society began approximately in 1990, then the era of the information society would be quite short; however, this might be an indication of the eagerness of Japanese government to position the advent of Society 5.0 as a major change.

There are some important points in the concept of Society 5.0. First, it is based on the Internet of Things (IoT) technology and the premise that societal issues can be solved by connecting various people with things as well as by sharing knowledge and information. Moreover, by fusing cyberspace and physical space, AI will analyze the data accumulated in cyberspace and provide feedback to the real world. Whether it is Industry 4.0 or Society 5.0, the important point is that various subsystems and devices now have the potential to be connected. Considering this under a technological perspective, this is a concept known as a "System of Systems," wherein individual systems work together to compose one substantial societal system. For example, electric power grid systems based on various aspects, such as RE, EV-centric networks of automobile companies, signal systems for autonomous driving, and communication infrastructure, are each designed separately. The implemented systems cooperate with each other, and this forms the electric power/transportation system of a city.

Alternatively, there is also the concept of "self-adaptive systems," wherein the system itself adjusts the functions and performance based on the situation. For example, this may include automatically controlled waiting times at a signal based on traffic, or dynamically adjusting the number of solar panels to incorporate into the grid, based on power consumption. Through this method, a system that automatically adapts to the environment is based on the underlining principle of this book: "effective use of resources." It also leads to the implementation of a system that provides what you need, when you need it, and only as much as you need.

In contrast, if these self-adaptive systems cooperate with each other to form a social system, mutual feedback regarding the adaptation status will become necessary, and these systems will become a complicated mechanism. The technological frontier in dissolution and reintegration is in collaboration and coordination; this is based on the premise that an autonomous adaptation of these independent subsystems is required. In addition, ensuring trust between subsystems that are designed and developed independently could be one of the important issues.

Open Networks for the Realization of Digital Twin

As an increasing number of devices and censors are embedded in the society, and the economic systems work as an interconnected network of functions, massive amount of data on social and economic activities can be generated and stored. These data make it possible to virtually reproduce the real world in cyber space, that is called "digital twin." Reality is recreated in cyberspace, and the results of the analysis conducted are fed back into the real world. One example in railway transportation is that, if location information is analyzed and an abnormal amount of congestion is expected at one station, a route using another station is automatically recommended.

For a complex service such as this, a wider range of reality must be pictured, compared to previous attempts. To this end, it is necessary to go beyond the conventionally "stovepiped" IoT and API ecosystems but these systems should openly cooperate. In addition, as we have observed in the System of Systems, it is crucial to have independent systems working together. The goal that the API economy and System of Systems is aiming to achieve is a wider range of collaboration. However, there is a notable barrier blocking this achievement. For example, until now, IoT has been stovepiped based on different businesses, such as the IoT of home appliance manufacturer A, that of automobile manufacturer B, that of telecom company C, and so on (Figure 3-6). The reason for this is that

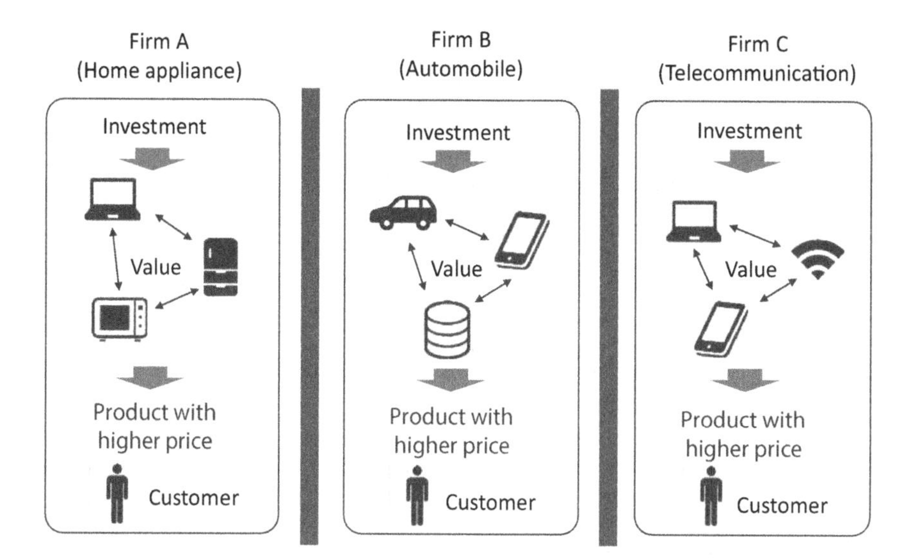

Figure 3-6 Stovepiped IoT ecosystem

it is difficult to earn a profit from an IoT service alone, thus the cost of building an IoT system must be recovered by including it in the price of the manufacturer's products. In a situation wherein businesses are competing for the superiority of their respective ecosystems, it is not easy to publish API data.

Therefore, by considering the exchange of data using API as a specific type of economic transaction and establishing a mechanism for capturing and exchanging the value of data, it may be possible to collaborate in a manner that transcends the "frame" of each business operator (Figure 3-7). This is a typical issue that blockchain, a fundamental technology behind Bitcoin, can solve, by symbolizing the value of data and making it transactable between parties, while avoiding the double-use and counterfeit. In particular, for protocols that exchange data using an API call, a payment procedure using digital tokens in the blockchain is incorporated, and a very small amount of digital tokens are paid in exchange for the data. A startup called 21.co (currently called Earn.com under Coinbase) has already implemented a mechanism to incorporate billing into APIs. If this type of mechanism can be realized, data can be distributed outside the "frame" of each business operator, and increasingly convenient services can be established.

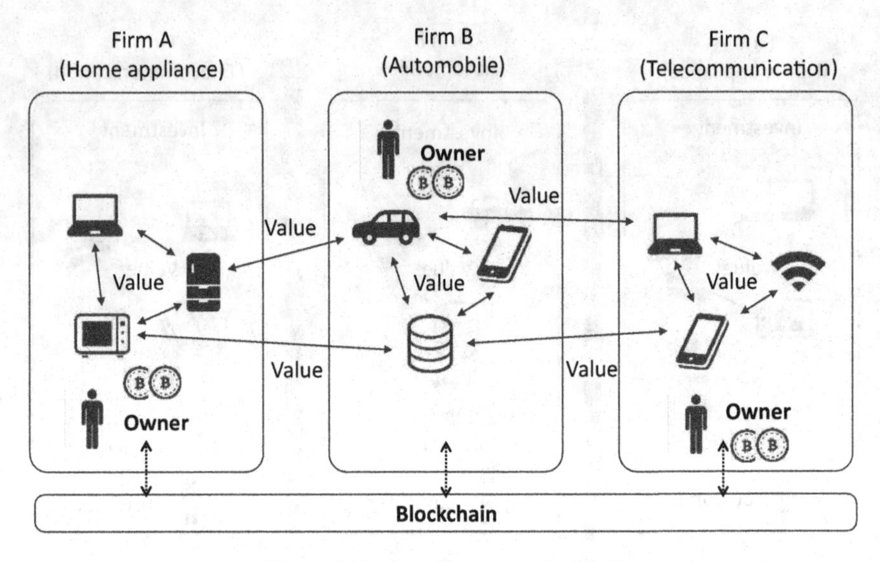

Figure 3-7 Seamless network of IoT

How Should a Business be Dissolved?

The first priority of a company should be selecting the component to deal with, from its traditionally packaged business domain. There are three criteria for evaluation. First, the potential size and seriousness of issues are a major factor. For example, in China, the use of mobile payments is rapidly spreading; however, online payment methods, such as electronic money and credit cards, have also existed for a while. In this case, why did QR code payments through smartphones become widespread? The reason is the cost issues on the retailer's side. In Japan, there are smart cards for transportation, but in addition to hardware costs for smart card readers, the transaction fee is approximately 3%, depending on the store. Similarly, there is also a 1%–5% charge for credit card transactions. Because this is borne by the retailer, these costs are invisible to the general consumer; however, we must assume these costs are reflected in the product price.

In contrast, QR code payment through a smartphone can be implemented simply by using a smartphone or tablet, thus a feature of this is the extremely low cost for implementation. In addition, transaction fees can be set at a cheaper rate or are even free, compared with conventional electronic money and credit cards. The use of mobile payments can rapidly spread if solves the retailer's problem of high costs of other online payment methods.

The second point is to acquire a substantial user base. Whether the business can expand from dissolved and specialized functions depends on the number of users in the first function that can be leveraged to achieve the economy of scope. In this sense, business functions that interest a large number of users could be a candidate of initial function. For example, communication, payment, commerce, entertainment, education, logistics, transportation, and food are the common interests of most people. Starting from these common features could be one of the options.

The third point is to select something that can introduce external resources, or be deployed over a platform-like service. For example, when offering educational content online, people who are knowledgeable in various fields can offer contents on various themes. For delivery, there must be a mechanism that allows external parties and organizations to provide delivery capabilities. Platform-like services are focused on the function of connecting external user groups, instead of having the resources required to directly provide the services. In the current era of deframing, the matching function becomes crucial in all settings to ensure a high possibility for horizontal expansion.

Dissolution and Reintegration as a Basic Strategy in the 21st Century

In this chapter, we have observed "dissolution and reintegration" as the foremost component of deframing. Providing a fresh perspective regarding traditional frameworks, such as banking, insurance, education, media, delivery, and transportation; extracting elements with potential for growth; and rebuilding using a digital-first approach is one of the most important perspectives while considering future business. As we have discussed in this chapter, even if a work is conducted focusing on a specified function, this does not indicate the end. It is then possible to expand the business through the use of economies of scope. "Dissolution and reintegration" is the starting point for innovation and the key factor for growth strategies in the digital age.

A business that grows through this method differs significantly from a traditional business because the business will be reconstructed completely differently, in comparison to the frames of traditional business domains. Business will be reconstructed in combinations that transcend the traditional framework, such as finance, communication, and manufacturing. In addition, the process of expansion requires the function of mediating, matching, and combining elements for users. This ensures the flexibility of values that can be moderated to serve for the user needs. This flexibility is called "specific-optimization," that is discussed in the next chapter.

Chapter 4

Specific-Optimization

The second component of deframing is "specific-optimization." Specific-optimization refers to optimizing for individual users that transcends the "frame" of ready-made products. In this chapter, we will consider changes in the business model that are caused by the evolution of technology, with specific-optimization as the keyword.

How to Respond to Diversified Values and Demands

In the past, most Japanese people were eating Japanese food, such as fish and miso-soup. However, as the economy expanded, people enjoy more diversified foods such as Italian, American, Chinese, Indian, and fusion cuisine. The values and demands of people are transforming from being single and uniform to become multiple and diverse.

Figure 4-1 shows selected items from chronological lifestyle survey on the Japanese people, provided by Hakuhodo Institute of Life and Living. These data show how traditional notions that were accepted as major values are declining in the last 20 years. For example, respondents who answered yes to "It is better to have more friends" declined from 57.2% in 1998 to 20.5% in 2018. Similarly, those who like Japanese food declined from 65.8% to 45.0%, and those who clean up their houses at the end of year declined from 68.5% to 56.1% within a year (in Japan, there is a traditional custom to clean the house at the end of year to prepare for a new year) in the same period.

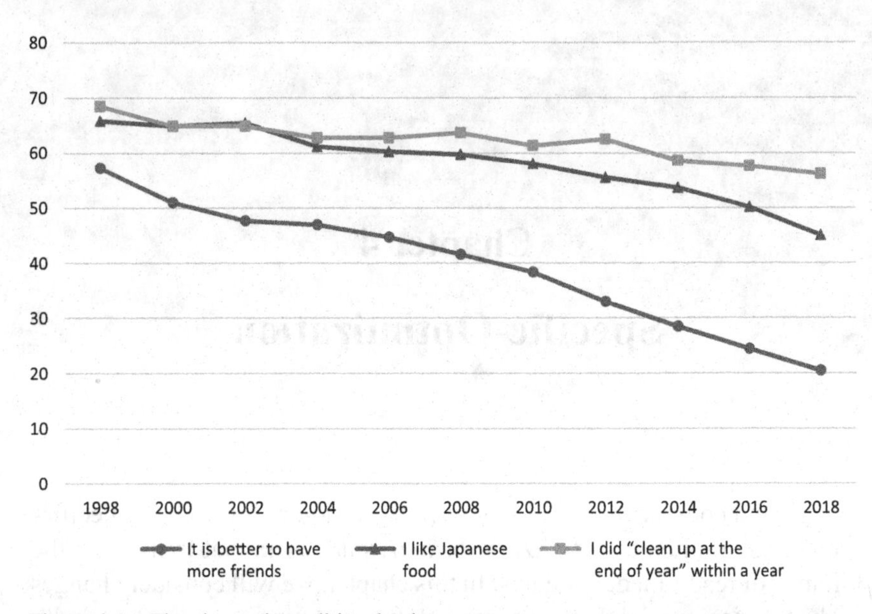

Figure 4-1 The change in traditional values

Source: Hakuhodo Institute of Life and Living, HAKUHODO Inc.

These data suggest how the dominant values and preferences of people are collapsing. In exchange, we have more diversified values and preferences in our life.

On the other hand, in the field of international economics, there is a concept known as the love of variety. This was advocated by a winner of the Nobel Prize in Economics, Paul Krugman. The concept is that the increased variety in a product or service ensures that it is more useful for consumers. This is a theory that highlights the advantages of living in a large city with diverse people, in addition to the value of trading with other countries that manufacture similar products. For example, in terms of cafes, there is an increased merit for consumers when there is not only Starbucks, but also Peet's, and Au Bon Pain. Instead of only having access to three types of domestic wine, it is better to be able to select from 10,000 types of wine, including imported ones.

First, we must recognize that fulfilling diverse needs is a significant benefit for consumers. Second, maintaining a certain scale without falling into a niche and understanding how to avoid an increase in costs becomes a crucial task.

Personalization of Information Services

The biggest challenge in optimization for individual users is the increased costs. One method of implementing specific-optimization while not sacrificing the economy of scale is to use the power of IT to minimize the customization costs and provide products and services featuring content that differs for each individual. The idea of customizing services for each customer has been developed in software-driven service sector, as the concept of "personalization." Fundamentally, personalization requires investment in a mechanism that uses information technologies, but once this has been done, the average cost for customization will decrease as the number of users increase.

For example, the method through which the search results are displayed and chosen by the users is personalized. When a user conducts a search using Google, the order of the search results may differ based on the user's details. This occurs because the search engine assumes "this person is probably looking for this type of information," based on the user's search tendencies and location information, and thus displays results that match the user. Based on the mechanisms, browsers have started to include a preload mechanism. Among the search results displayed in the list, the page that the person is most likely to click on is downloaded in advance, and when it is clicked, the page is displayed instantly, which allows for faster browsing. In addition, online advertising has been personalized since the last decade. The Google AdWords mechanism analyzes a user's search history and displays advertisements the user may be interested in. Once that mechanism is created, the user's input becomes valuable source for analysis, and the service is further optimized for that user. News websites have also been personalized. For example, Yahoo Japan started personalization of Yahoo! News in 2013, so that they can automatically display the contents that match the interest of users. This kind of personalization is widely adopted in various information services, such as YouTube, Amazon Prime Video, and Netflix.

On the other hand, there is a critique that the personalization to this extent while searching for information is a major issue. When displaying only information that matches a user's preferences, such as political beliefs, it will appear to the person that the world is full of people with similar beliefs. Eli Pariser called it a "filter bubble," which makes it impossible for users to understand information from the world in a well-balanced manner.[1] It also leads to users engaging in an “echo chamber” when discussion becomes unidirectional, because the person begins to feel as if they are surrounded by those with the same opinions. It is noteworthy that information retrieval, a service that is seemingly uniform, can be provided as a service that is optimized for each individual.

Personalization that Transforms Manufacturing into Service

Mass customization does not stop with information services. Since 2012, Nike has been offering a service called Nike ID that allows customers to combine the color scheme and size of sports shoes. Customers are able to freely select and order from a large number of color combinations online and that information is sent to the Nike factory, after which customized shoes are delivered to the customer. Moreover, the cost is closely similar to a ready-made product. I have used this service several times myself and this has helped me save time by ordering directly, instead of going around searching for shoes in my preferred size and color scheme.

This service type has become possible because of multiple factors: the interface arrangement that allows users to configure what they want and place an order online, the system that manages the order information and sends it to the factory, and the manufacturing system that allows for immediate manufacturing, based on the order details. Similar customization systems are offered by Cannondale and Trek, which sell sports bicycles. These allow customers to flexibly select from a vast number of colors and equipment to create their own bicycle.

In another case, ZOZO Town, an e-commerce platform for fashion, had once provided “ZOZO suits” that enables customized T-shirts and suits for customers. It provides a tight fit uniform with dot pattern for customers to measure the exact size and shape of their body. Customers wear the uniform and take a picture of themselves and send it to a server, then the service analyzes the change in dot patterns to analyze the accurate size and shape of the body. Based on this analytics, ZOZO suits produces the T-shirts or suits that best fit the customers’ body.

---

[1] TED. https://www.ted.com/talks/eli_pariser_beware_online_filter_bubbles?utm_campaign=tedspread&utm_medium=referral&utm_source=tedcomshare.

Another interesting example is provided by Shiseido, a cosmetics company. Shiseido provides "Optune," a personalization service of basic skincare products for customers.[2] It produces the best-fit skin lotion for each day from a combination of five ingredient cartridges that are located in customers' place, based on the analysis on a variety of information such as the skin condition, external factors (temperature, ultraviolet light, humidity, pollen, etc.), and internal elements (biorhythm, stress, mental condition, etc.). Basically they sensor the skin condition from an application and camera on smartphones and utilize image recognition. This is a subscription model that costs 10,000 JPY monthly, providing a personalized service for skincare. This is enabled by the combination of technologies of sensors in smartphones, analytics on environmental data, and manufacturing technology for customization.

Similar service is also proposed by Johnson & Johnson as "Neutrogena MaskiD."[3] It provides customized face-pack for skincare based on the analysis of the condition of skin that is sensed by cameras on smartphones. It aims to analyze the skin condition of the face through a detailed mesh and put the proper skincare lotion in the proper place on the face-mask.

As seen in these examples, personalization requires the combination of various technological developments, such as sensors, manufacturing, and logistics. However, the most notable advancement in the last decade would be the integration of sensors and analytics technology, such as that provided by smartphone cameras and image analysis empowered by neural-network artificial intelligence (AI). Image recognition and analytics has become an extremely important method for converting various types of information around us into data and utilize them in various aspects in our society. The cost of collecting necessary data for customization has significantly decreased, compared to that of the past.

This type of customization is also emerging in the field of hardware, a more traditional manufacturing sector. TeamLab's Masakazu Takasu has written a detailed report about this in *The Maker's Ecosystem*,[4] regarding the ecosystem of "makers" (people who develop their own hardware,

---

[2] Shiseido. https://www.shiseido.co.jp/optune/#system. Additional reference: Horiki, S. (2019). Diversity of personalization in cosmetics market, *Japan Marketing Academy Conference Proceedings* 8 (original in Japanese, translated by the author).

[3] Neutrogena. https://youtu.be/VnwdLXHZeNw.

[4] Takasu, M. (2016). "NicoNico Engineering Department Shenzhen Observation Meeting." In *The Makers Ecosystem*, Impress, Tokyo.

including for their own hobbies) in Shenzhen, China, and the company Seeed is introduced at this instance. Seeed produces electronic circuit boards in very small lots to fulfill the needs of "makers." Customers worldwide can use the Internet to upload data, the board is produced in just a few days and sent to them. There are other hardware and circuit board manufacturers, such as Elecrow, that can support small lot orders.

In the past, it is typically difficult to enter into production without the assurance of being able to sell several tens of thousands of pieces for the development of the hardware. However, because the exchange of design data and payment systems has been significantly streamlined by IT, manufacturing circuit boards in units of ten has become a possibility. Furthermore, with the permission of the person who placed the order, Seeed publishes the design data to allow other users to place an order using the same blueprint, or customize and modify it. This is a strategy that allows for scaling by making it open source, while simultaneously supporting specific-optimization for each user's needs.[5]

This concept of not only rolling out something that has been individually optimized but also using it to create an ecosystem or community that includes other users, will likely become increasingly important. Because mass customization involves data exchange, this indicates that the manufacturer will accumulate a significant amount of data regarding customer preferences. Using this data for marketing and deploying it across related services are crucial.

Beyond Servitization

As seen in the previous section, not only software-driven service sector but also manufacturing sector has been combined with the elements of services. Traditionally, the combination of manufacturing and services is called as "servitization." A research group of the University of Cambridge defines servitization as "the process of creating value by adding services to products. It involves offering "fuller market packages or 'bundles' of customer focused combinations of goods, services, support, self-service and knowledge in order to add value to core product offerings.""[6]

---

[5] *Ibid.*

[6] Dinges, V., F. Urmetzer, V. Martinez, M. Zaki, and A. Neely (2015). *The Future of Servitization: Technologies That Will Make a Difference*. Cambridge Service Alliance, University of Cambridge.

Certainly, in many manufacturing industries, a major issue is planning methods for breaking out of the business of simply producing and selling hardware, and planning methods aimed at increasing added value by providing services associated with the hardware; this is done by utilizing the data generated from using the hardware. On the other hand, personalization of manufacturing is not merely “adding services to products,” but transforms the nature of products into services. Therefore, let us go back to the importance of services and look at how it is important for the future of manufacturing.

The importance of the service industry does not need to be emphasized yet again; however, it is approximately 70% of the GDP on average in developed countries, or based on country, it is 77% in the US and slightly lower in Japan, at 69%.[7] In general, the proportion of the manufacturing industry tends to decrease as a country develops because, as the manufactured products become commoditized, the only differentiating factors are costs, such as labor and land costs; therefore, the industry moves to countries where costs are as low as possible. For example, garment production has moved from the US to Japan, from Japan to China, and more recently to Cambodia and Vietnam. The production of items, such as steel and electronics, will also be transferred to developing countries. In contrast, developed countries become service-focused, by creating added value through services that are difficult to commoditize.

This is also true for individual companies; for example, among home appliance manufacturers, general products (such as washing machines, refrigerators, and televisions) are exposed to competition from developing countries with cheaper land and labor costs. Currently, they have managed to survive by relocating their factories to developing countries; however, it has become difficult for them to compete with the growing number of manufacturers that have emerged from the same developing countries. This is currently the situation for Japanese home appliance manufacturers.

In these circumstances, various moves have been made to include services to hardware products. This ranges from the traditional, such as maintenance, financial services for purchases, and insurance, to increasingly varied services, such as navigation support in automobiles and providing knowledge obtained from analyzing the usage.

---

[7] *Source*: World Bank, 2016 values. https://data.worldbank.org/indicator/NV.SRV.TOTL.ZS.

Professor Cusumano of the Massachusetts Institute of Technology[8] has cited the software industry from the 1990s as an example, which transformed from a business specializing in software packaged by the industry, to one that combines packaged software and additional services (such as consulting, customization, and maintenance). The reason for this, Professor Cusumano highlighted, is that it is possible that the price of these software products becomes zero. Recently, the Windows operating system (OS) has become free, and Google offers software for free that has similar functions as that of Microsoft Office. Thus, we can observe that the expectation of Professor Cusumano was significant, based on the fact that various platforms, such as Facebook and Slack, are provided for free.

Whether it is hardware provided by the manufacturing industry or packaged software provided by the software industry, those that are packaged to suit everyone will eventually become commoditized and their value will reach zero. In the case of the manufacturing industry, there are inevitable costs for raw materials; therefore, it will not be free. However, among those with an extremely low marginal cost for copies, such as software, there is a possibility that the price will reach zero. For this reason, it is important to prevent commoditization and protect the interests of companies by converting products into services, in addition to combining products and services.

What is a service and what are its characteristics? In addition, why is it possible to prevent commoditization by converting something into a service? In general, a service has four characteristics in contrast to that of hardware[9]:

- We are unable to touch it (Intangibility).
- The content differs based on the customer (Heterogeneity).
- Production and consumption happen at the same time (Simultaneous production and consumption).
- Services cannot be stored (Perishability).

For example, let us consider the service typically provided by a hairdresser, i.e. to cut hair. It is impossible to touch the act of cutting hair.

---

[8] Cusumano, M. A. (2010). *Staying Power: Six Enduring Principles for Managing Strategy & Innovation in an Uncertain World.* Oxford University Press.

[9] Zeithaml, V. A. and M. J. Bitner (2003). *Services Marketing: Integrating Customer Focus Across the Firm.* 3rd edition, McGraw-Hill, p. 21.

We can touch the hairdresser, but the hairdressers themselves are not the service. In addition, the details regarding the cut differ entirely, based on the quality of the customer's hair and their preferences. The customer is consuming the service at the same time when the hairdresser is creating the haircut. In other words, production and consumption occur simultaneously. The act of cutting hair cannot be stored somewhere and it disappears immediately.

This is the traditional definition of a service, and these conditions apply to various aspects, such as consulting, cooking, construction, and education. There are products that are derived as a result of these services, but these properties are present in the actions that are taken. Due to this nature, the provider and customer must "cooperate" to create the service. In other words, a service is intrinsically customized and individually optimized.

However, there may be a question that is raised here. Are the services provided by Google and Apple considered uniform even if each customer is different and the user does not participate in the production of any services? Certainly, early cloud-based services did not have any customized components and only provided the functions provided in turn by packaged software through the network. However, as the number of users increased, abundant data regarding customer preferences and usage was accumulated, and technologies analyzing data, such as data science and AI, became widespread. Therefore, minutely customizing these services to fulfill the needs of customers is no longer an impossible task. This is the modern customization of cloud services and mass customization technology.

As mentioned earlier, digital technologies are transforming the manufacturing sector into the services sector, not only adding services to products. This is made possible economically by the development of technology and also by the motivation to exit from commoditization.

The Back-and-Forth History of Specific-Optimization

Particularly, the IT industry has experienced swinging trends between packaging and specific-optimization. Figure 4-2 shows whether the industry is aiming for general optimization by packaging elements to uniform products or services, or specific-optimization that is tailored service for unique customers.

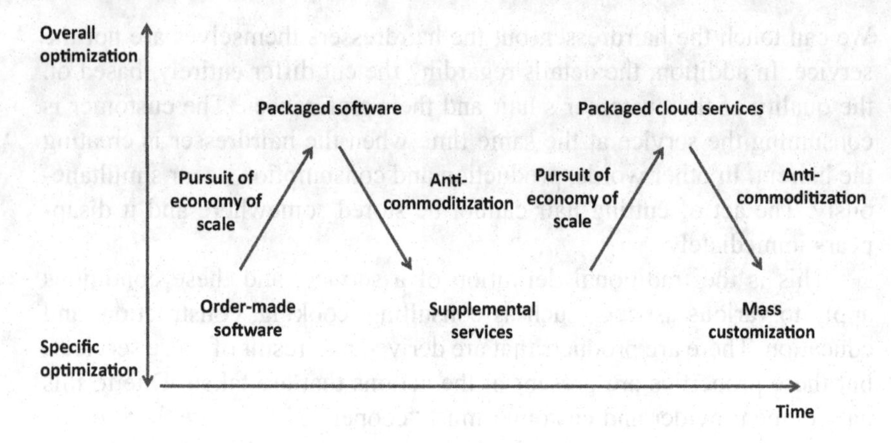

Figure 4-2 Swinging trend in software industry

Since 1940 when electronic computers were born to the early 1970s, software was completely custom-made to suit the customer's demands.[10] For example, the UNIVAC which was developed in 1951 for the census and later used to predict the presidential election as promotional purposes, was necessary to custom-create numerous programs for this purpose. In this era of customization, the scale of the business was undoubtedly small. IBM is known for a claim of Thomas J. Watson Sr., that "there was a market for no more than about a dozen computers and that IBM had no place in that business."[11] This was around 1949, and it paints a picture of how the computer was a niche product. However, IBM changed its strategy in 1951 and began to work on computers whose roles were substantially changed from numerical calculation to data processing, and within five years, they were dominating the market.[12] Computers began to be mass-produced as industrial products, similar to the Ford Model T.

Nevertheless, the era of custom-made software continues. The airline seat reservation system (1963) was also developed by IBM for American Airlines. However, during the 1950s and 1960s, programming languages, such as FORTRAN, COBOL, and BASIC, were created and anyone who studied these could develop software. Simultaneously, the price of

---

[10]Campbell-Kelly, M., W. Aspray, N. Ensmenger, and J. R. Yost (2019). *Computer: A History of the Information Machine*, 3rd edition, Routledge.

[11]*Ibid.*, p. 97.

[12]*Ibid.*

hardware began to decrease, thus making it possible for the general individuals to purchase a computer.

When an explosive growth in the number of computer units was expected, it was Bill Gates, the founder of Microsoft, who created the business of distributing large quantities of packaged software. He started a business that earned US$10–50 each time a PC was sold with an OS called MS-DOS, which Microsoft had developed at that time.[13] Software took off from the era of customization and burst into the "era of packaged software," which pursued economies of scale. Packaged software, such as Microsoft's Windows, Apple's Macintosh, Office software, Oracle, and SAP, can be duplicated at a marginal cost of nearly zero, which has led to substantial profits for these companies.

However, the golden age of packaged software did not last for long. As the market matured and competitors emerged, either prices sharply reduced or the software became free. An open-source OS, Linux, that is comparable to other OS, such as UNIX, was soon released. In addition, open-source database products, such PostgreSQL and MySQL, which are comparable to Oracle and Microsoft SQL Server, were also released. Recently, functions similar to those provided by the Microsoft Office product line are being offered by Google for free through cloud services. This suggests that packaged software is facing the challenge of commoditization. As a result of open-source software, significant profits from the mass sales of packaged software and price destruction may seem like a contradiction at first glance; however, it is the result of both extreme of the same phenomenon that software can be copied at nearly zero marginal cost, and the characteristic of "information" that indicates friction is substantially low with relation to distribution.

In response to this commoditization, each company has strengthened its service division. Since then, IBM focused on its consulting and systems integration business. Moreover, they sold their notebook computer business called ThinkPad, which had tremendous brand power during that time, to Chinese company, Lenovo. Similarly, Microsoft is focusing on its cloud business with Azure, in addition to a business that is close to system integration.

Each company, in opposition to the commoditization of its packaged software, has adopted a strategy that makes it difficult for customers to easily switch to another company or free software. They did so by

---

[13] *Ibid.*, p. 247.

providing customer-oriented services and added value that is indispensable for customers. The flow of the software industry in the 20th century has shifted from specific-optimization to packaging, and then back to specific-optimization.

In this century, we have begun the second lap of the zigzag race between packaging and specific-optimization. "Platform" services have emerged, including Google's search and online advertisement, social networking services (SNS) such as Twitter and Facebook, and Apple's iTunes. These represent a type of cloud computing that provides pre-packaged functions over a network, but instead of being limited to their own functions, they depend on the functions and information provided by others, and this differs from conventional packaged software because the specialization is in the matching function. However, the early versions of these services feature uniform content and this has been a service that relies on scale.

The platform has its own specific network effects; thus, it will not be commoditized immediately, but development and implementation costs are not notably expensive and switching can easily occur. Facebook is not the only SNS; LinkedIn, and in recent times, China's WeChat have gained prominence. In Japan, Mixi was at its peak in the 2000s, but in the blink of an eye, users switched to Facebook. The ideas regarding SNS functions and the technology used to achieve these services can spread quickly worldwide. If there is something that is slightly better or that differentiates it from the rest, it will spread easily and users may switch from the previously used platform.

Companies offering these packaged cloud services and platforms must be prepared for commoditization. It will become necessary to provide services that are closely customized to each user. By fine-tuning the functions to match user needs and understanding user preferences and circumstances to provide an appropriate service, companies will want to create a situation wherein that user cannot easily switch to a different platform.

Looking back at the history of the software industry since its advent, we have experienced a zigzag pattern between packaging and specific-optimization at least twice thus far, and we can see that we are currently in the second period of specific-optimization. The pursuit of specific-optimization has become important as a response to commoditization, including the resulting maturity of the market and price decrease.

Will this zigzag pattern between packaging and specific-optimization be repeated in the future? Although this possibility persists, there is also another perspective in modern society. That is, people's values and way of life are diversifying and there is a universal need for services to respond to individual circumstances. Therefore, specific-optimization is still in demand, not only from an economic rationale but also based on larger social trends.

Achieving Variety Through the Use of a Platform

In addition to personalizing services to fulfill customer needs and manufacturing customized products, another method for achieving specific-optimization exists. This method is matching by platform, which leverages a massive number of various third-party providers. This makes it possible to offer a wide variety without having to directly provide the products and services.

For example, in the field of education, there are various platforms, such as Udemy, Schoo, and Dedao. These platforms offer a diverse range of educational content, including web development, illustration, design, foreign languages, investment management, and entrepreneurship. Udemy has over 35,000 instructors registered to teach courses and offers 80,000 online courses (as of January 8, 2019). Each type of content may fulfill unique user needs. Moreover, if the instructor is an individual, they can receive sufficient income. This type of specific-optimization cannot be fulfilled by a large corporation that expects to sell a substantial volume of products to the mass market. In addition, because the platform does not have its own content, it can prepare a variety of products that fulfill a diverse set of needs.

This applies to food delivery platforms, such as Ele.me and Meituan, which are popular in China. Through these apps, it is possible to find not only Chinese food from various regions, such as Beijing, Shanghai, and Sichuan, but also other types of cuisines, such as Vietnamese, Thai, and Japanese.

Ride-hailing applications, such as Uber and China's DiDi (Didi Chuxing), also offer various vehicle models and services. For example, DiDi has a varied menu, such as services for a single rider, carpool, luxury car, airport only, charter, traveling to Hong Kong (only in Shenzhen), six-seater car, traditional taxi, bicycle, designated driver service, and reserved car. This level of variety would be difficult to offer for usual taxi companies, with the exception of significantly large companies.

"Economy of scope" is crucial for achieving specific-optimization in the platform era. This is because a wide range of components must be prepared to provide the combination of components required by customers; however, if everything is done in-house, it would be unviable from a cost perspective. By relying on external resources, it is possible to flexibly provide a variety of clothes, food, housing, a workplace, and transportation methods that suit the customer's tastes.

If Something is Truly Necessary, It is Worth the Cost

Through this method, utilizing information technology is a key to reduce the cost of customization while pursuing the market of specific-optimization. However, even it is not digital-intensive services, sometimes customers accept the increased costs to access tailor-made services that help them to achieve their personal goals.

Rizap is a company that pursues this type of specific-optimization through personalized training in the fitness industry. The concept is to create a program that suits the individual's current situation and goals on their fitness and training, and to closely follow the customer until the goal is achieved. Although costs are incurred, users are willing to bear these costs if they can achieve their goals (e.g. weight loss or increased muscle strength). Recently, Rizap has expanded beyond fitness to advance into the areas of golf and foreign language learning. Its fees are relatively higher; twice a week through two-month training costs 298,000 JPY,[14] significantly higher than usual gyms that cost around 11,000 JPY per month. Nevertheless, its service is quite popular.

In other case, a company called Noom uses a smartphone app and AI to provide optimized health advice for its users. Additionally, a registered dietitian offers individual advice regarding various aspects, such as diet and exercise. The service is thoroughly tailored to individual user needs.

Sitateru, an apparel-related service based in Kumamoto, coordinates the free-time of garment factories spread all over Japan to provide a service that produces clothes in small lots for individual orders. It can be considered a platform-like company that produces products that fulfill individual needs and are not mass produced by linking a substantial number of factories.

[14]RIZAP. https://www.rizap.jp/course/course_detail, as of December 24, 2020.

The goal of specific-optimization is related with the outcome, and not the output. Until now, among packaged services, importance was placed on the following question: "What do we have to offer?" It was sufficient to offer a specific service, whether it was fitness or English lessons. It was up to the users to use the service for achieving their goals. In the era of specific-optimization, the focus is on the outcome (i.e. the end result or goal). Specific-optimization is necessary for producing the outcome because each individual user has their own set of circumstances and different goals. Simply providing the same service will not ensure that the desired results are achieved. It goes without saying that what the user truly wants is not only to use the service, but to reach the state they are aiming for, and that is where their perceived value of the service lies. Focusing on specific outcomes naturally requires specific-optimization.

Specific-Optimization for Hardware

Thus far, we have discussed specific-optimization for consumer products or services, but similarly, the pursuit of specific-optimization is becoming important in areas in hardware, infrastructure that is invisible to the general users, and also in regional governance. In the following section, I will discuss the technological and social aspects involved.

Inefficiencies in overall optimization have become also apparent in the hardware industry. John Hennessy, an early developer of the CPU, the President of Stanford University and Chairman of Google, advocated for the development of CPUs specialized for different applications at Google's developer festival "Google I/O" (conducted in May 2018), and referred to this as "domain-specific architecture."

Hennessy's points are as follows. One of the important points was the end of Moore's Law, which states that the degree of integration on an integrated circuit doubles every two years, thus indicating that computer performance will improve exponentially. There is also Dennard's Law, which states that the amount of power consumed by one unit of silicon remains constant. Until now, Moore's Law and Dennard's Law have been applied to the design of integrated circuits. Although the degree of integration on a piece of silicon continues to increase, power consumption is constant; therefore, processing capacity per unit of energy has improved dramatically.

Due to the spread of smartphones and IoT, there has been a dramatic increase in the number of computers used globally. In this context, it would be a problem on the imbalance of energy supply and demand if Moore's Law does not continue. Unfortunately, Moore's Law is beginning to break down. Hennessy showed that it has taken seven years, not two, to double CPU performance in the current era.

Therefore, the issue is the creation of a computer that reduces power consumption and the key to solving the issue is the development of a CPU specialized for its application. Until now, CPUs have been developed with the aim of executing various commands at the same time for any application. Therefore, although they are versatile, their power efficiency is poor. By developing a CPU specialized for its application, such as AI and virtual reality, the idea is to create a power-efficient computer. Until now, attempts have been made to create software to improve the performance of general-purpose CPUs; however, these perform poorly in comparison to dedicated hardware. Hennessy showed, that in comparison to a program written in Python, processing using domain-specific architecture is 62,806 times faster.[15]

An example of specifically optimized hardware is a chip called application-specific integrated circuit (ASIC), and there is one of them specialized for Bitcoin mining. Bitcoin mining requires a substantial volume of special calculation called hash processing. Based on the speed of this calculation, it is possible to receive bitcoin as a reward for mining. Although hash processing is possible using general laptop computers and smartphones, in terms of competing for speed on a global scale, having dedicated hardware is overwhelmingly advantageous. ASIC chips are manufactured in significant quantities in China and the reason behind the existence of this individually optimized hardware was that China was once the center of the mining industry.

As Hennessy stated, because power constraints have been revealed as well as from the perspective of using resources without generating waste, specific-optimization is in demand. Of course, there are costs involved in producing dedicated hardware, and new designs as well as investment in manufacturing equipment is required. Therefore, an important perspective of innovation in the future is focusing on the reduction of the costs of design and manufacturing for specific-optimization and customization, including hardware.

---

[15]The future of computing: A conversation with John Hennessy (Google I/O '18). https://www.youtube.com/watch?v=Azt8Nc-mtKM.

Specific-Optimization of Infrastructure

By ensuring the full use of new technology, such as sensors, payments, and data analytics and AI, infrastructural services can also be optimized, thus making it possible to use resources more efficiently.

For example, in Singapore, electronic road pricing (ERP) was implemented, wherein the billed amount is optimized according to the value of each resource on various roads around the city, such as degree of traffic congestion, time frame, and lanes.[16] Each person's needs from a road are different. There are people who are willing to pay more to go faster and those who are willing to travel slower and save money. Existing services and infrastructure can be transformed to optimize for any situation by integrated with ICT services.

Two of the important technologies here are sensor technology and authentication technology. If it is impossible to confirm which car passed through, a billing omission may occur, or another car may be charged twice. One reliable method of authentication is the use of an in-vehicle device and a gate, such as with Electronic Toll Collection system. However, in recent years, another powerful method has emerged — image recognition.

Amazon is using this method in the US for its pilot cashier-less supermarket, Amazon Go. This method recognizes each customer's purchases through the analysis of facial recognition and images of their movements. The payment is then charged to the credit card of the person identified by the authentication information. This eliminates the use of a gate wherein people "stand in line to pay," and thus allows people to seamlessly buy items.

In addition, a Chinese company is providing technology that makes gas and water meters smarter by retrofitting parts. A micro-camera reads the numbers on the meter and sends this information to the center by using a mobile phone network to centralize the data. This eliminates the need for staff to personally visit and read the meter. Through the analysis of this data, it may be possible to discover and respond to various user needs. Therefore, meters do not need to be replaced entirely to make them smarter.

Meanwhile, parking meters that recognize license plates have recently emerged in Japan. In regular parking meters in Japan, a flap is used that

---

[16]Takasu, M. (2016). "NicoNico Engineering Department Shenzhen Observation Meeting." *The Makers Ecosystem*, Impress.

stands upright to prevent the car from moving without payment. The flap is lowered only after the payment is made and the car can then be moved from the parking lot. In contrast, lots that use license plate recognition do not include a parking flap. The user simply pays and leaves. There is no barrier stopping someone from not paying; therefore, it is possible to leave without paying. However, because the license plate has been recognized and recorded using a camera, the user will be billed for this during their next visit to the parking lot. This method allows the driver to park without any stress, because there is no forced mechanism using a machine that sometimes cause mechanical trouble and damage on the cars. It also can reduce installation and maintenance costs for the equipment.

Transforming the physical activity to digital data, and feeding back the result of analytics into the physical world are the indispensable elements of changing infrastructure into specific-optimized services, and new technologies such as image recognition can lower the hurdle to implement such a digitalization.

Deframing in Regional Revitalization

The principle of deframing also applies to social areas, such as regional revitalization. An essential aspect of regional revitalization is determining the characteristics and values that completely differ across regions, because there is no optimal solution that applies to all regions. In this context, virtual currency or blockchain technology can be used as a technology that can convert values characteristic to a region into visible forms available, thus contribute to solve the specific challenge of the region.

The fundamental difference between virtual currency based on blockchain technology and traditional money is that the value of virtual currency is not backed up by a specific, trusted organization. The novelty of virtual currency is that the payment record is managed by an unspecified number of participants, and the algorithm prevents its counterfeit or double-spending payment. It has become possible to create a medium (generally called money) for recording and distributing value, despite not being backed up by an authoritative organization such as central banks.

As a result, it became possible to create an economic sphere that can flexibly adjust while remaining loosely connected to the world of fiat money. For example, in Aizu-wakamatsu city, in northern part of Japan, an experiment was conducted to capture the value of community and to liquidate it in November 2016. This experiment regards a communication between residents as a value for the community, and designed the system to issue a newly created money of "Moeka" in exchange with the action of meeting other people and conversing with them. This led to many young people having conversations with others to ensure they could receive the money. In the experiment, Moeka could be used to buy official goods of Moeka and other things such as coffee.[17]

If the premise was to meet up to receive a certain amount of fiat money such as 100 Japanese yen, the value would inevitably be compared to that of 100 yen in the usual life where the Japanese yen is used as currency, and the participants would be more hesitant. By issuing a currency that is focused on the social value of the encounter, it is possible that an economy with completely different standards of value can be established. This could be also applicable for utilizing mediums for exchange for the data in IoT or contribution within universities or clubs.

By providing a method of recording value using blockchain technology, it is possible to reveal the value exchanged in a specific society, and further promote the circulation of the value. Without being tied to a single currency such as the Japanese yen, it is possible to communicate with others using a value form that is suitable for each economic zone. While discussing virtual currencies, emphasis is placed on negative aspects such as speculative actions and hacking, but its significance remains in revealing various values and making them tradable.

Issues with Overall Optimization in the Euro

While observing specific-optimization from a very broad perspective, applying one currency to a wide variety of economies is not without its challenges. The Euro, introduced in the EU, is a straightforward manifestation of this challenge.

---

[17]Takagi, S., H. Tanaka, M. Takemiya, and Y. Fujii (2017). "Blockchain-Based Digital Currencies for Community Building," *GLOCOM Discussion Paper Series* 6 (17-004). http://www.glocom.ac.jp/wp-content/uploads/2017/06/GLOCOMDISCUSSION PAPER_No6_2017No.4.pdf.

As political and economic integration deepened in Europe, the Euro was created in 1999 as a part of the efforts to compete with economies of scale, such as the US, particularly through the formation of a single market. The intended, direct solution using the single currency was to reduce exchange costs in the region and the risk of exchange rate fluctuations. With a unified currency, there would be no need to consider exchange costs or risks, thus making it possible to freely trade and invest in the region. The expectation was that optimal allocation of resources would be promoted by utilizing economies of scale on a wider range.

In contrast, to participate in the Euro, it is necessary to fulfill the following requirements: price stability, low long-term interest rates, exchange rate stability, and fiscal discipline (fiscal deficit-to-GDP ratio is below 3% and accumulated debt-to-GDP ratio is below 60%).[18] These restrictions have led to the secondary effect of improving public finances of each country in the region and stabilizing interest rates and prices.

However, there are also negative effects of the single currency. Prior to the introduction of the Euro, countries could lower interest rates to encourage lending during times of economic downturn, and it could result in currency depreciation, which would stimulate exports. In addition, if each country has its own unique currency, when imports significantly exceed exports, sales of the home currency increase and the exchange rate depreciates, this results in imports becoming expensive and exports becoming cheap. Through this method the balance between imports and exports is automatically adjusted. Joseph E. Stiglitz, a Nobel Laureate in Economics, has pointed out that the introduction of the single currency, the Euro, made it impossible for European governments to make use of these economic adjustment mechanisms, thus resulting in significant trade imbalances among the EU countries.[19]

For a single currency to function without these negative effects, in addition to having very similar economic conditions across the region, the four types of free mobility must exist — free mobility of products, services, capital, and people. The mobility of people is important because of the idea that if there is an economic problem in a particular area, then moving to an area with increased economic power will solve the problem. The US dollar functions in the federal system of the US as a single currency because of the free mobility of people.

---

[18] Tanaka, S. (2010). *Euro: The Unified Currency in Crisis*, Iwanami Shinsho.

[19] Stiglitz, J. E. (2016). *The Euro: And its Threat to the Future of Europe*, Allen Lane.

However, although freedom of mobility exists in Europe, there are certain limits to migration across countries, due to the influence of each country's system, culture, and language. Based on the fact that a single currency does not allow financial adjustment according to the circumstances of individual regions as well as restrictions on people's freedom of mobility, mass unemployment can occur in some countries, as is the case of Greece.

The reduction in currency exchange costs and the risk of exchange rate fluctuations was originally cited as a benefit of introducing a single currency in Europe. Even general tourists can sufficiently understand the time and cost of exchanging money every time they cross a border and the loss on coins that can no longer be used. In addition, if one were to make investments across regions, and if a risk of exchange rate fluctuations exists, there will also be risks to business performance and cash management.

On the other hand, depending on the use of technologies, it would be possible to reduce risks by linking individually optimized currencies. For example, due to the spread of credit and debit cards in recent years, there are significantly less opportunities for using cash while traveling internationally. In China, mobile payments using smartphones that read QR codes to make payments are widespread and are used in various places, from subway ticket machines to convenience stores, taxis, and greengrocers. If only the necessary amount is processed online and automatically exchanged and settled, even if a foreign exchange commission is charged, the user will save more time.

In addition, for the foreign exchange commission, the ask price (TTS) and bid price (TTB) were set at the time of exchange, and the difference was the profit of the exchange office, which was a cost incurred by the user. However, a venture company called TransferWise has succeeded in using only domestic remittances to effectively provide overseas remittances, by matching the demands for funds in each country, instead of directly transferring money overseas, thereby making it possible to exchange money at the middle rate (TTM). If this system becomes increasingly widespread and advanced, not only individuals but also companies will be able to conduct business without being conscious of exchange costs.

As mentioned earlier, although currency fluctuations are a risk, they also act as a control valve on trade imbalances. If exchange rate fluctuations are a barrier for corporate investment and sales, methods aimed at actively removing this barrier can be considered. A typical method would be currency hedge transactions using derivatives. Until now, this has been primarily handled by large companies, institutional investors, and financial institutions; however, the situation would change if more user-friendly product can be developed that even general consumers can use for day-to-day economic activities. Foreign exchange is fundamentally zero-sum, and if the interests of market participants can be adjusted, it may be possible to mutually hedge fluctuation risk.

In Europe, as noted in the issues of Catalan independence, there are two forces at work: integration under the large umbrella typically seen as the EU and the independence of the micro community. Kenichi Ohmae, the current President of Business Breakthrough University, proposed the "region state" that should be a natural economic unit of an appropriate scale.[20] A future issue regarding the socio-economic system will be related to the use of the power of ICT to balance global efficiency against the diversity of economic transactions that take advantage of the circumstances and characteristics of individual regions.

Limitations of Individual Optimization

In this chapter, we have observed the second component of deframing; specific-optimization. It is necessary to respond to the true needs and circumstances of the user. As shown in the example of shoes, a person's needs can be fulfilled more accurately if a product or service is tailormade to suit them, instead of forcing users to search for a product or service that closely resembles the aspects they are looking for. Advances in IT have made this level of customization possible without incurring significantly high costs.

However, customization and personalization are not necessarily all-purpose. Sometimes proposing a goods as a package of values that the customer is not aware may yield an improved result, compared to if the customers design it themselves. The role of a designer is to present a solution based on their visual and modeling skills to solve any type of problem. It is likely that they will suggest something far better than the usual customer would design.

---

[20]Ohmae, K. (1995). *The End of the Nation State: The Rise of Regional Economies*, Kodansha (in Japanese).

Moreover, in some cases, the primary motivation for purchasing items, such as branded products, may be to receive approval from others. Instead of having a product that suits the user and only has value to the user, it may be more beneficial to the user to have a product which many people place high value on, for example, a Hermes Birkin bag. Furthermore, considering the resale value, it may be easier to sell to second-hand dealer if it is a ready-made product.

Therefore, packaging products based on existing conventions may be the more rational choice for products wherein the user does not have a strong preference, for branded products wherein other people's evaluation is important, and for items wherein resale value is important. However, for all other items, products that match user needs as much as possible can be provided, thus leading to increased user satisfaction, and the continuous use of the product can be expected.

Specific-optimization is also a differentiating strategy to avoid commoditization and to transform products into services. It is also the most efficient use of resources, depending on the demand, because it is a response to diversifying individual tastes and needs. To enable specific-optimization, it is necessary to combine various technological factors, such as mass customization, personalization, sensors that convert information into data, analytics technology including AI, and platforms that enable economy of scope. The challenge for the future is to create a business and socio-economic system that can fulfill individual demands and support a diversified economy.

Chapter 5

Individualization

In this chapter, we will look at the third principle of deframing, "Individualization." Individualization refers to the process by which individuals play more active roles, going beyond the "frame" of the corporate organizations. This is an element that will affect future workstyles, career design, and learning styles.

Individuals Have the Resources for Modern Business

From the perspective of essential resources for modern businesses, what has changed the most is who has the information. Prior to the Internet, the extent of information transmission was limited, so it was extremely difficult to gain an understanding of the entire picture and latest trends of a business. It was not possible to have complete knowledge except for someone at the head of an organization employing many people to collect information or paying a consultant a large sum of money. However, we are now in an era in which, for the most part, anyone can access the world's most advanced information and participate in communities on the Web. It is now possible for new employees and students to gather cutting-edge information from various areas of society without having to aggregate information through organizational strength. While information has become an important resource for management, it is no longer the case that the higher up one is in an organization, the richer the information one holds. Similarly, IT skills such as programming can now be easily acquired, as the latest information is distributed with increasing speed. In addition, due to technological advances, it is now possible for 1 person to create an application that 100 people were once necessary to create.

Organizations are also said to be "information processing machines." As shown in the section on theory in Chapter 2, in the past, to ensure efficient decision-making, information was aggregated from on-site employees and sent to the top of the hierarchical organization. In this model, efficient management was carried out by making decisions at the upper levels. In this case, it was common to "extract" and reduce the volume of information so that it could be effectively digested as it moved to the upper levels. Communication channels that transmitted this information would then define the culture of the organization (Figure 5-1).

However, in the modern business environment, there are limits to such information-dense, hierarchical organization management. First, the technical information that is needed for the business changes so quickly that it becomes impossible to keep up with changes in the environment if management waits for the information to come up from below. It can happen, though it may be an extreme example, that even if employees in the field realize the importance of a new technology, it may take months or years for this information to reach the president.

Second, the essence of the latest trends, particularly for IT, may not be properly conveyed if the information is extracted and reduced. There is a big difference in one's understanding between reading about the existence of something called "a sharing economy" in a report of paper, and actually using Uber or Airbnb. If information is extracted, this can lead to the missing of important points that include the essence. Instead, it is better to actually see and touch the real case to understand deeply.

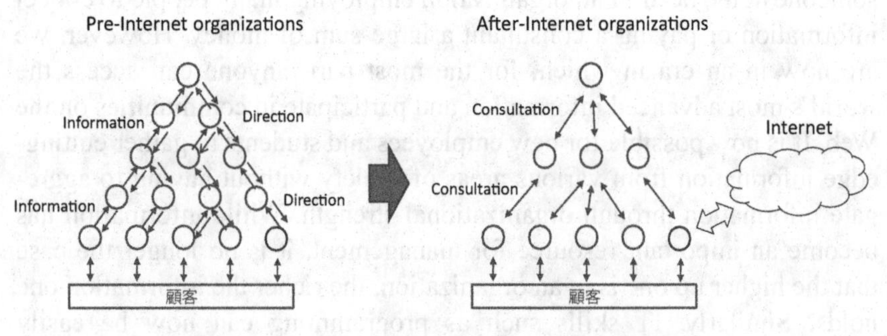

Figure 5-1 Organization and information processing

Third, in the past, business targets were clear, and once a communication channel was established to gather the information needed for decision-making, it could be used for a long time. For example, in the semiconductor business, it would have been sufficient to collect information on a regular basis about global semiconductor prices, the product development trends of major manufacturers, construction, and operation of factories. However, if the target business changes frequently, the communication channel that was once established may no longer work. In an era of rapid changes, communication channels must be as short and flexible as possible.

In other words, rather than letting people who do not have the information make a collective decision, entrusting control as much as possible to people who have the necessary information and skills will lead to the effective functioning of the organization and the efficient utilization of resources. At the same time, this change in access to information has become a major driving force for the individualization.

The Sharing Economy as an Epitome of Individualization

A simple embodiment of individualization is the services of the sharing economy. From service providers' perspective, a network of drivers who have empty seats in their cars and spare time has significantly reduced the costs associated with traditional employment. It became possible for a person who holds the resources that meet a need to perform a task in demand. Uber successfully identified the needs of both drivers and riders and acquired a market within the red ocean of taxi services. From the customer's perspective, Uber, which offers a ride-sharing service, is the almost the same as a taxi service. However, payment with Uber is simple because the customer registers their credit card in advance, it is easy to request a ride using GPS on a smartphone, the destination is inputted in advance so there is no need to explain the route, and the route is recorded, which reduces the risk of detours. Thus, all the hidden complaints related to taxi services have been eliminated, and the service has been widely accepted by the market. The service is not often seen in Japan due to regulations, but along with similar services such as Lyft and DiDi, it is widespread in many countries such as the US, the UK, and China.

As seen in ride-sharing, most services that appear to be new are actually transformations of existing services. Crowdfunding, for example, opened up and enabled the visualization of the traditional mechanism of donations, allowing anyone to contribute to a project. Up to now, there were many complicated procedures related to donations. It was difficult for a donor to determine where the donation funds were going, or sometimes because of lack of clear procedures, the donor had to go through the trouble of contacting the counterpart to coordinate how to make a donation. The power of IT has dramatically improved the mechanism of matching and payment, making it easier to find the beneficiary of the donation.

In addition, the simplification of the procedures has greatly expanded the domain of donations. As a result, individuals, shops, and restaurants that had never before thought of asking for donations have been introduced to the idea of collecting donations through crowdfunding. The important point here is that, as a result, a new task has been created, for example, the produce of attractive crowdfunding projects. Even if the essential function of business has not changed, when transaction costs have been significantly reduced, it results in qualitative changes and more business variations.

After the outbreak of COVID-19, crowdfunding is gaining more popularity for the purpose of supporting those who are suffering from restrictions of operation, such as restaurants and inns. For example, READYFOR, a crowdfunding platform, started the special initiative to support various aspects of the economy such as medical institutions, childcare, pro-sports, restaurants, and artists.[1] It mediated more than 840 million JPY support from 20,577 donors until July 30, 2020.[2] According to Mitsubishi UFJ Research and Consulting,[3] the market size of crowdfunding in Japan is 132.66 billion JPY (around US$1.288 billion[4]) in 2019. However, lending-type crowdfunding constitutes 86.8%, and purchase and donation type constitute only 12.7%, whereas it is growing from 7.7 billion JPY in 2017 to 16.9 billion JPY.

---

[1] READYFOR. https://covid19.readyfor.jp/#project-list.

[2] Mitsubishi UFJ Research and Consulting, September 30, 2020. https://www.caa.go.jp/policies/policy/consumer_policy/caution/internet/assets/caution_internet_201013_0001.pdf.

[3] Mitsubishi UFJ Research and Consulting, September 30, 2020. https://www.caa.go.jp/policies/policy/consumer_policy/caution/internet/assets/caution_internet_201013_0001.pdf.

[4] Calculated with US$1 = 103 JPY.

Increasing Number of Freelancers

Freelancing symbolizes the deframing of workstyle. Put simply, freelancers are self-employed workers who are not employed by companies but conduct work through a contract with customers such as companies to earn a living. As previously mentioned, the practice is growing in Japan, but this shift has been more widely seen in the US.

In the US, new technologies are adopted quickly, and the employment system is more liquid than other countries, so these changes tend to happen quickly. For example, the impact of offshore outsourcing of tasks such as software development and call centers on employment was already a political issue in 2004,[5] and ride-sharing has firmly been established.

In recent years, freelancing has gradually increased across the US. According to a survey by Edelman Intelligence,[6] 57.3 million people are freelancers in the US, and this number increases every year. In addition, the ratio to the total working population has reached 35.8% (Figure 5-2). Freelancers' contribution to the US economy is said to be US$1.4 trillion,

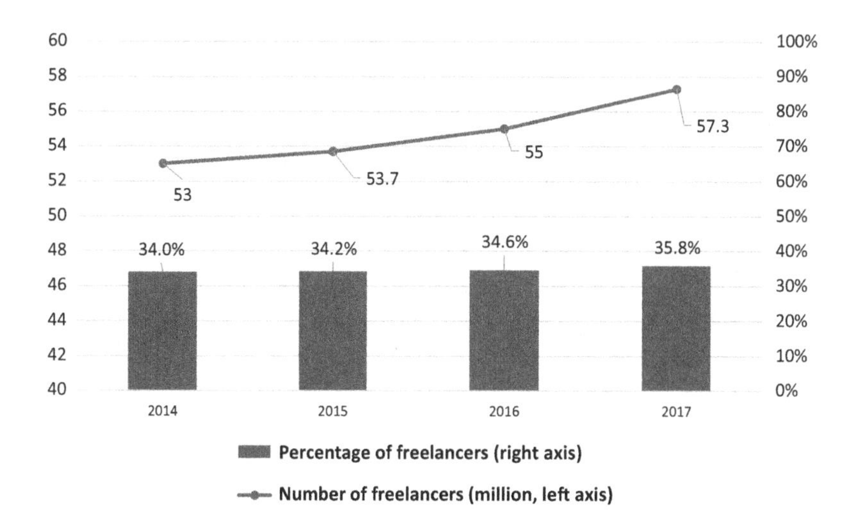

Figure 5-2 Trend of freelancing

Source: Edelman Intelligence (2017).

[5]Mankiw, N. G. and P. Swagel (2006). *The Politics and Economics of Offshore Outsourcing*, https://scholar.harvard.edu/files/mankiw/files/outsourcing_-_march_7_2006.pdf.

[6]Edelman Intelligence (2017). *Freelancing in America: 2017*, https://www.slideshare.net/upwork/freelancing-in-america-2017.

and as the US GDP is US$18.5 trillion, this means that about one-tenth of the economy is supported by freelancers. This scale is large enough not to be ignored. However, even though their ratio to the working population exceeds 30%, freelancers' contribution to the GDP is about 10%, and thus it should be noted that that freelancing is not always financially advantageous.

The same survey suggests that there are many freelancers in the younger generations, rising to 47% of the working millennial generation. Therefore, if these generations and beyond continue to choose to freelance, there will be an even higher percentage of freelancers. The survey estimates that the number of freelancers will surpass employed workers in 2027.

Respondents to the survey stated that they chose to become full-time freelancers because they want "to be my own boss," that is, to have autonomy in their work, as the top reason, in addition "to have flexibility in my schedule," "to be able to choose my own projects," and "to work from the location of my choosing." In other words, there is an increasing demand to work freely with greater autonomy and flexibility (Figure 5-3).

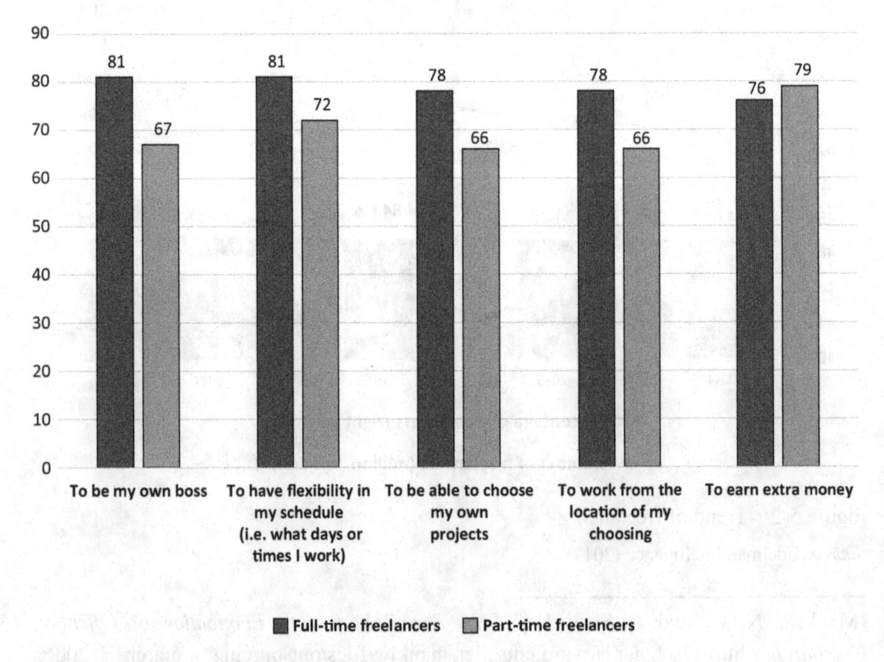

Figure 5-3 Reasons for taking up freelancing

Source: Edelman Intelligence (2017).

The number of freelancers in Japan is also increasing significantly, reaching 10.87 million workers in 2019.[7] These movements mark a change in peoples' values regarding their way of life, and is the result of inefficiencies where information-empowered individuals are working in hierarchical organizations, such as large corporations.

Individualization of Industries Driven by Platforms

The existence of platforms that connect individuals and clients has been essential in enabling people to perform work independently. As previously discussed, transporters are now the key to services in China, delivering everything from meals to items from the convenience store and supermarket. These transporters participate in the market through platforms such as Meituan.

The individualization of industries through platforms is evident in crowdsourcing, which matches individuals with a wide variety of tasks such as software development, website creation, design, and article writing. Multiple platforms are offered in Japan, such as CrowdWorks and Lancers. Figure 5-4 shows the scale of the transaction value of crowdsourcing in Japan, which is expected to grow to the scale of 300 billion JPY in FY 2020.[8] As seen in Chapter 1, the percentage of employees who want to take part in side jobs increased from 4.4% in 1992 to 6.4% in 2017.[9] If second jobs/side jobs become more popular in the future, the number of users of crowdsourcing may further increase.

Individualization due to advancements of platforms can also be seen in more specialized areas. In the field of education, one of the leading examples of this is the platform "Udemy." Udemy originated in the US and provides more than 100,000 courses that are mainly useful for improving the skills of working adults in fields such as programming, business, design, and marketing. Among these is a course through which one can comprehensively learn programming (HTML5, WordPress, CSS3,

---

[7]NIKKEI. https://r.nikkei.com/article/DGXMZO60710750T20C20A6EA2000?type=my#oAAUgjQwMA.

[8]*Source*: Yano Research Institute "Result of a survey on crowdsourcing market, 2014, 2016" https://www.yanoict.com/summary/show/id/330; Ministry of Internal Affairs and Communications "2015 White Paper on Information and Communications in Japan, https://www.soumu.go.jp/johotsusintokei/whitepaper/ja/h27/pdf/index.html" A document of Crowdsourcing association.

[9]Statistics Bureau of Japan. https://www.stat.go.jp/data/shugyou/2017/pdf/kgaiyou.pdf.

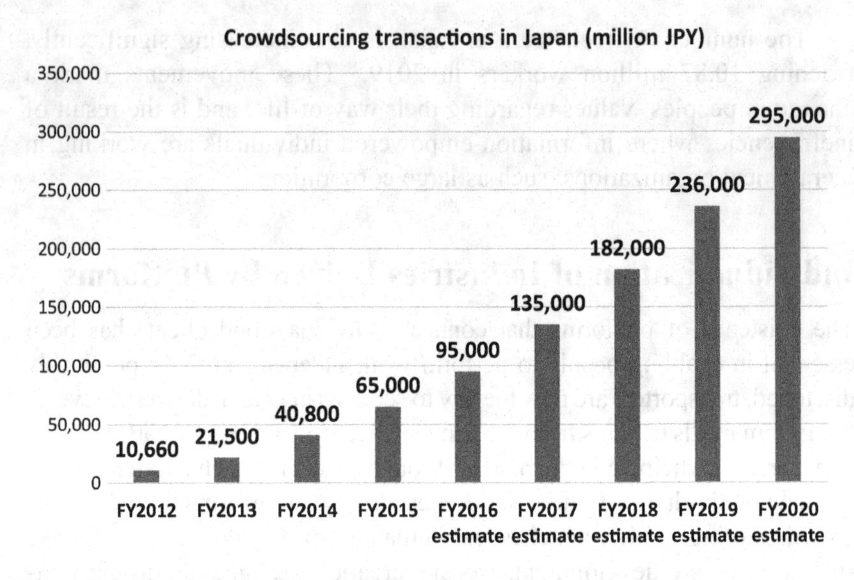

Figure 5-4 Trend of crowdsourcing transactions
Source: Yano Research Institute; Statistics Bureau of Japan. https://www.stat.go.jp/data/shugyou/2017/pdf/kgaiyou.pdf.

JavaScript, Bootstrap 4, PHP, Python, MySQL) and become a web engineer. There is a course that about 300,000 people worldwide are enrolled.[10]

Anybody can enroll in a course that costs anywhere between US$10 to several hundred dollars. In addition, more than 35,000 instructors around the world are registered to teach on the platform and earn income each time a student purchases a paid course. Instructors can work as individuals to perform educational work through Udemy, and students can enroll in individual courses.

The individualization of this type of education is also progressing in China, where paying for knowledge is commonplace and an "online paid knowledge market" has been established. The scale of this market is expanding every year, and it is estimated that it is around US$750 million 2017 and will reach about US$3.6 billion in 2020.[11] Multiple platforms mediate the "online paid knowledge market" in China, one example of which is an app called "Dedao." With this single smartphone app, it is

[10]As of the publication of the Japanese edition of this book.

[11]iResearch (2018). *China Online Knowledge Paid Market Research Report*. (Original in Chinese); Calculated with Chinese Yuan = 0.1548 USD at January 7, 2021.

possible to listen to lectures in a range of fields such as self-development, technology and innovation, humanities, philosophy, sociology, history, economics, finance, business, commerce, life sciences, physics, astronomy, and medicine. Each course consists of several to hundreds of lectures, each of which lasts 10–30 minutes. The lectures are generally in audio format, but it is also possible to read transcripts. This allows users to study during small gaps of time, such as during their commutes. When I visited China, I met a young woman working in the technology sector who also used Dedao to study.

Experts in a variety of fields publish their lectures on Dedao. One course can cost 200 Chinese yuan (about US$30), and because in some cases there are about 200,000 subscribers, the sales would be US$6 million in a simple calculation. According to an iResearch report,[12] some factors behind the popularity of services for which people pay for knowledge is the popularity of mobile payments and various platforms, which have made it easier to discover and pay for these services, and that people have formed a habit of paying for valuable content.

Until now, the education market existed in the framework of schools and universities, where students attended pre-determined lectures, but this has begun to change dramatically. With the spread of platforms, we are entering an era in which we can learn what we need, as much as we need, and from teachers that we like. In the future, learners may be more likely to take initiative in designing and enrolling in their own educational curriculum, and managing their own educational experiences. In this case, it would be useful to have blockchain technology that can record and verify this without being dependent on a specific educational institution.

For example, Blockcerts, developed by MIT Media Lab and Learning Machine, is a blockchain-based credential service for certificate. It can store certificates for civic records, academic credentials, professional licenses, workforce development, etc., and provide a function that a reviewer of the certificate, such as an employer, can verify it by checking the blockchain.[13] The University of Nicosia also uses a blockchain to prove that a student completed a particular course by storing the information of certificate onto the blockchain.[14] These services can enable to build

[12] *Ibid.*

[13] Blockcerts. https://www.blockcerts.org/guide/.

[14] University of Nicosia. https://dcurrency.wpengine.com/wp-content/uploads/2015/11/dfin511-index1-final.pdf.

an original certificate that consists of flexible combination of the courses provided by various institutions such as universities, schools, and companies.

Influencer Marketing by Individuals

Meanwhile, one important perspective from a business organization is how to interact with and utilize these individual actors. For example, in media sector, people with important knowledge, awareness of the issues, insight, and the ability to comment on a specific issue are not limited to the persons in in the media sector, such as television stations and newspapers; someone who are engaged in different jobs in different fields may be familiar with the background of individual news and events, and others may be able to provide rich commentary. NewsPicks is a service that makes good use of these people's capability of commentary and points of view.

The NewsPicks service publishes articles from the Internet selected by users (pickers) with comments added. Since it is also linked with various social media, it is possible to add one's own analysis to the news as a commentator and allow it to be seen widely on the Internet.

The reader can discover how people who are active in various fields interpret certain news and how they have the opinion on the topics. With a variety of pickers, it is possible to be exposed to a variety of opinions, rather than only those of TV commentators. In addition, this meets the needs of the "pickers" who want to express their views about current affairs, and thereby increase their presence in the market. Particularly active pickers are appointed by the company as a "Pro Picker," who are paid to post comments on the news.

On the other hand, the business model for YouTubers is to offer various video content on YouTube, the video distribution system operated by Google, in exchange for advertising revenue. By continuously uploading videos of about 10 minutes in length, from product introductions to fashion, golf lessons, and funny videos, they can expect to generate a fixed audience and continuously earn advertising revenue. YouTube has more than 2 billion users and more than 500 hours of content are uploaded every minutes. Particularly after the COVID-19 pandemic, YouTube has established itself as a major presence as a media with a variety of content ranging from yoga, comedy, and hobby to art performance. This could be partly driven by the viewers, who have to stay home to avoid infection, and also creators who lost job opportunities, particularly in the entertainment and art sectors.

The YouTuber phenomenon symbolizes individualization, in that it is possible to produce programs and develop a business similar to a traditional TV station without having to be a celebrity or own specialized equipment. However, as will be discussed in detail in Chapter 7, as competition between YouTubers intensified, they gradually became more specialized. Nowadays, there is a management company that specialize in YouTubers; thus, it is important to note that not everyone is able to make a living from it.

In other case, influencer marketing points to the mechanism of individualization in the field of advertising. Even without being a celebrity, by posting one's own fashion or trends on social networking services such as Instagram, the user can acquire a large number of followers. When these ordinary people post about items such as clothes, food, and drinks on social networking services, they can offer significant advertising possibility. By utilizing this potential, the practice through which companies hire contents creators to post about the company's products to raise awareness about the products is called "influencer marketing." For influencers, if the product is actually stylish or something that can make viewers to admire them, it will improve their power of influencing. Viewers can also benefit from learning how to use the product through influencers' demonstrations.

However, generating posts that the general public would not recognize as an advertisement can be misleading for consumers; it is thus undesirable and is called "stealth marketing." Recently, requirement has emerged for the use of hashtags such as #sponsored to clearly indicate when posts are sponsored. The Word of Mouth (WOM) Marketing Association, which was established with the aim of developing WOM marketing in Japan, revised its guidelines in December 2017 and requested that social networking services (SNS) posts in which a product or fund has been provided by a company clearly state this support. It has thus become necessary to clearly indicate that the context is different from what normal consumer may assume by using tags such as #promotion or #sponsored. The market of influencer marketing is expanding, and its size in Japan is 31.7 billion JPY in 2020, which was a 5% increase from the previous year. Almost 40% of them is on YouTube, followed by Instagram and Twitter, according to a survey.[15] Similar case is seen in China as *Wanghong*; The

[15]A survey by CyberBuzz and Digital InFact, in NIKKEI. https://www.nikkei.com/article/DGXLRSP541559_S0A011C2000000/.

3.5-million-strong *Wanghong* are ordinary people who are trying to promote and sell goods on social media platforms.[16]

With the emergence of information and communication technology (ICT) services like social media, the practice of advertising carried out by individuals has become feasible, instead of using celebrities in a large-scale mechanism in terms of broadcasting and communication power such as conventional television commercials. The most important point here is that, in the case of influencer marketing, advertisements are effectively delivered only to those who are following a specific influencer and who are interested in specific topics, compared to mass advertising such as traditional TV commercials. According to a case study conducted by a marketing company called Nielsen Catalina Solutions, when 258 fitness and food influencers carried out a campaign called "Meatless Monday," the return on investment was 11 times higher than that of conventional mass advertising. Pinpointing where users actively look for information is the second element of deframing, i.e. specific-optimization.

New Workstyles Supported by Co-Working Spaces

Until now, we lived in an era in which we belonged to a frame called the "company" and only needed to do the work our boss told us to do. However, in present times, the need has arisen to create new services that have not existed before. If we stay tethered to the desk and environment given to us by the company, there are limits to any new awareness we can acquire.

On the other hand, when individuals have more resources available to them than before and are able to work independently and proactively, there is a risk that the individual becomes isolated. Although it may be easy to access information, it is not possible for one person to become an expert in everything. For new ideas to emerge, it is important to interact with others with different knowledge and skills. As scholars on economic and technological development, such as Joseph Schumpeter and Brian Arthur, expressed that innovation emerges from the new combination of factors,[17] innovation is also born from a combination of existing

---

[16] *NHK World*. 2019. Chinese Internet celebrities: Influencing the huge online market.

[17] Schumpeter, J. A. (1983). *The Theory of Economic Development*, Routledge; and Arthur, W. B. (2009). *The Nature of Technology: What it is and How it Evolves*, Free Press.

technologies, and it is very important to interact with people with experience in different technologies and knowledge.

In this context, co-working spaces are becoming important as organizations that support interactions between these kinds of individual actors. There are various names for this, such as shared offices, co-working spaces, or incubation facilities, in which various people such as startup company founders, freelancers, and company employees can share spaces and work together.

One study[18] found that the supply of flexible workspaces are increasing on average 16% in major cities worldwide, and London, New York, Hong Kong, Paris, Shanghai are hosting the largest supply of those spaces. This growth is especially remarkable in London; WeWork, a major co-working space operator rents the largest space of offices in London, apart from government agencies.[19]

The author conducted a field survey of co-working spaces in London in November 2018, and it is observed that various types of spaces have been created — from hotel lobbies that are used as co-working spaces and major co-working spaces such as WeWork, to the spaces that are characterized by a sense of community and relaxation. WeWork was operating 482 co-working spaces in 96 cities around the world at the time of the survey. There are about 40 in London alone, and the number is said to be increasing by one each month. The interiors of WeWork are all unified to a certain degree, with a "free address" open space for everyone to use, a kitchen for free drinks, and private office spaces for rental by corporations. There are many glass partitions, and the office is designed with basically industrial tastes. In addition, there is always a draft beer tap in the kitchen, symbolizing that this is a space in which you can interact with others with a beer in one hand.

In addition to freelancers, startups, and small businesses, some departments from large companies occupy units at WeWork. Workers are attracted to the ability to easily use a functional and comfortable office space that has been superbly designed. Such spaces are also advantageous for startups in their attempts to recruit human resources. In addition, they make it possible to respond flexibly when the business

---

[18]The Instant Group. The Global Flex Market: The top 18 markets for Flexible Workspace in 2019.

[19]*Financial Times*. https://www.ft.com/content/40a87044-ff97-11e7-9650-9c0ad2d7c5b5.

scale expands, such as in the case of startups. Furthermore, if a worker is to visit another city or country, they can use the same chain's co-working space, allowing them to be in their accustomed working environment.

However, the selling point of this type of co-working space is not only its stylish design and flexibility. WeWork, for example, offers a dedicated application to find information about other tenants. The application can be used to search for others who specialize in web design or database experts, for example, and find their locations, and quickly place work orders. Instead of interacting with people from the same company every day, it is stimulating to have daily contact with people working for different companies, industries, and contexts. In other words, a co-working space is not simply a "shared office" but a new community and labor market in place of a company. Rather than distributing work based on hierarchical commands, this new organization offers adaptability in work among people who share values related to workstyle on the premise that the entity is more individual-driven.

The growth of these co-working spaces is the result of the changing values of workstyles. Until now, the primary model has been to join a large company and succeed while climbing the career ladder. However, the reality was that only a limited number of people would succeed in this model, and only a limited number of people who joined a well-known company could participate in the race in the first place. In comparison, since the beginning of the 21st century, startups have been founded through investments with their ideas and visions, and young people have responded to the workstyle of building assets in the form of an Initial Public Offering (IPO) or the sale of the company.

In reality, few companies reach the IPO stage, and many people leave startups to work for other companies. However, the values have partly changed regarding workstyle, from loyalty to a large organization which requires waiting for gradual advancement based on a boss's evaluation, to an emphasis on the vision and product, receiving direct feedback from the market, and collecting capital gains. This means that the workstyle implemented by the tech industry in the last several decades has become the norm.

Against the backdrop of this type of workstyle, an increasing number of people place importance on interacting intellectually with other like-minded people in an office designed to stimulate creativity, rather than working in cubicles as a member of a large, homogenous organization.

Providing a Variety of Co-Working Spaces

These types of co-working spaces are also gradually increasing in Japan. For example, the Yahoo Japan Corporation operates a co-working space, "LODGE," which uses 1,330 square meters of space in a building in a prime location in Nagata-cho, at central Tokyo. It is currently free of charge for anyone to use, so people at various occupations work side-by-side at their desks. According to the company, the space is used by about 320 people every day, and their occupations vary from individual business owners, to company employees and startups. In addition to desks designed for various purposes, the space is fully equipped with a corner for food and beverage and Wi-Fi. In addition, at LODGE, a staff member called a communicator speaks to users to grasp their interests, in addition to conducting initiatives that facilitate connections between users, and regularly holds tea parties. In offering this much space for free, the income and expenditure alone might not balance positively, but Yahoo! Japan has another motive; their purpose is to create collaborations between people with various skills who gather here and Yahoo! Japan employees. For example, in one case, a user of LODGE with 3D skills and a Yahoo! employee created a service that visualizes the aspects of electronic commerce.

Meanwhile, as a contrast to LODGE, there is a co-working space called "Office Camp Higashiyoshino" in Higashiyoshino Village, Nara Prefecture, which has a population of only 1,700. Designers and others gather here in the renovated space of a traditional Japanese house. A company called Office Camp LLC was created from this hub, where individuals interact and create new frameworks. Incidentally, all of the people in the company moved here from other areas, so in order to achieve a balance between a lifestyle in the countryside and being able to do the work they want to do, the co-working space fulfills the role of being a place to find "colleagues in the countryside." In co-working spaces, the function of encouraging and supporting individual actors is becoming increasingly important.

7F (*Nanaefu*), which is a one-minute walk from Omiya Station in Saitama City, Saitama Prefecture in Japan, is also unique. Most of the workers here are entrepreneurs, but the space is characterized by services designed with corporations in mind, such as corporate registration, address usage, fixed lockers, and mailbox services. In addition to these services, importance is placed on communication between users, and there are events such as book launch parties to encourage interaction between users, and also mentoring for entrepreneurs. GaiaX Co. Ltd. also operates a co-working space and attempts to create interactions between users by frequently holding events. These mechanisms are important because it can be difficult for strangers to spontaneously start a conversation. In addition, GaiaX employees use the co-working space, and in some cases, go to other co-working spaces or work from home, providing a very flexible workstyle, which could even be called an extreme workstyle.

In case of Japan, because of the shy national character, it is necessary to work more proactively to create interactions. According to Dr. Tuukka Toivonen at the University of the Arts London, who studied co-working spaces and creativity, there are various types of co-working spaces, such as those focusing on exchange and collaboration, and specialized co-working spaces for specific industries such as fashion. In some cases, co-working spaces simply meet peoples' needs to have a space where they can focus on their work. Alternatively, the needs may change from day to day: people may want to interact one day and concentrate another day. Dr. Toivonen is developing a "radar chart" that can help users find a co-working space that suits them.[20] As workstyles become more diverse and the number of jobs people can do as individuals increases, it is becoming an important issue to develop infrastructure that enables new connections that are different from those within conventional companies.

Industrial Promotion Should Focus on Individuals

In the past in China, state-owned oil companies and telecommunications companies used to represent the majority of employment providers, but as China shifts to major entrepreneurial country, startups have been extremely active, with many platforms emerging to support them.

There are several thousand co-working spaces throughout China. For example, "XNODE," a co-working space in Shanghai, targets company founders and combines co-working with a startup accelerator. Various companies are based here, including freelancers, companies with 3 to 4 people, companies with 30 to 40 people, as well as investment companies, allowing investors to immediately invest in promising entrepreneurs. Additionally, to operate XNODE itself as a creative space in which people

---

[20]Toivonen, T. (2018). A multivitamin for the future of work? Rethinking the value of coworking for digital creatives. *RSA Future of Work Blog*, November 7.

can solve problems together, there are various mechanisms in place to facilitate interaction through such as welcoming parties and events. When I happened to visit XNODE, a study session on blockchain was being planned.

According to Weilin Zhao of Fujitsu Research Institute, who is an expert on Chinese startups, in Chuangye Street (Founder Street) in Zhongguancun in Beijing, entrepreneurs are supported with the funds, technology, and personal connections necessary for starting a business. This neighborhood is home to the unique "garage cafe," which was established by an angel investor in November 2011. In Hangzhou, which is known for its hosting of Alibaba's headquarters, there is a region commonly known as "Dream Town," which is a platform for corporate support that aims to prevent high-skilled employees who have left Alibaba from leaving the area, and instead encouraging them to start new businesses in Hangzhou. China's economy has grown from a manufacturing center that had business outsourced from overseas, to an "entrepreneurial paradise" in which young entrepreneurs bring about new innovations every day. Those who fail to understand this dynamism and only follow large, traditional companies will certainly be left out of the global business flow.

This kind of entrepreneurial culture is emerging in Japan as well. Figure 5-5 shows the number and amount of venture investment in Japan

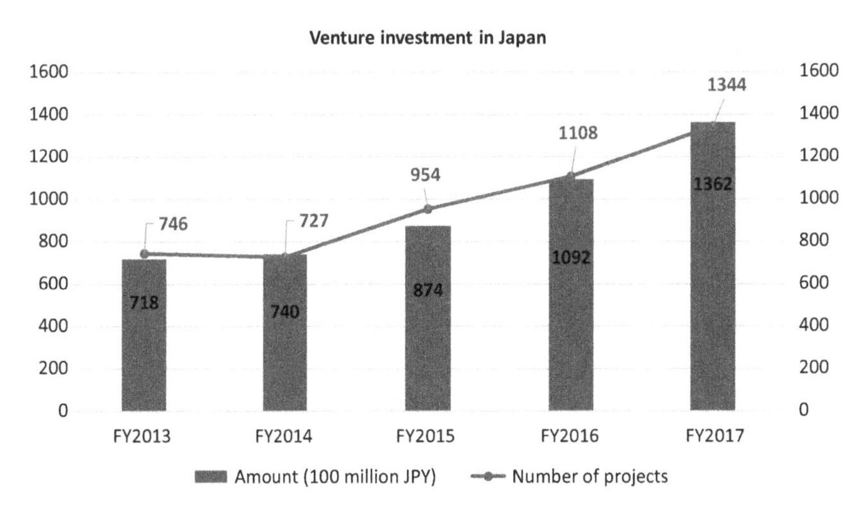

Figure 5-5 Trend of venture investment in Japan

since 2013, and it is on an increasing trend as a whole. The entrepreneurial boom came once in the early 2000s, and the economy is now recovering from the cooling caused by the subsequent 2008 financial crisis. However, in recent years, the ecosystem of accelerators, startup pitch contests, and investments by large companies in venture companies known as Corporate Venture Capital (CVC) has been enhanced, and the base of entrepreneurs has expanded significantly compared to the past.

Now, with the power of ICT, starting a business has become relatively easy. Industrial promotion in the 20th century was typically represented by large-scale investment and attracting factories, but going forward, it is becoming more important to support the establishment of businesses that are driven by individuals.

Part III

Organizations and Individuals in the Age of Deframing

Chapter 6

Trust in the Deframing Society

Thus far, we have seen how technological advances have dissolved and reintegrated elements of services, and how individuals instead of organizations have emerged as key players in the economy. In fact, economic changes due to deframing, particularly the third element individualization, are deeply related to the issue of trust. In this chapter, we will consider what "trust" means in the first place, why "trust" is important in the age of deframing, and how we can enhance and secure trust with evolving technologies.

Increasing Importance of Trust in the Deframing Age

What does trust mean in the first place? The meaning of "trust" differs depending on the field of study. According to a survey by Professor Rousseau at Carnegie Mellon University, economists see trust as computable, while psychologists consider it to be an internal cognitive problem. Further, sociologists understand it as a characteristic embedded in society. Although there is no commonly accepted definition of "trust" in every field, the professor and her colleagues state that the widely held definition of trust across disciplines is the following.[1]

[1] Rousseau, D. M., S. B. Sitkin, R. S. Burt, and C. Camerer (1998). Not so different after all: A cross-discipline view of trust. *The Academy of Management Review* 23(3): 393–404.

> *Trust is a psychological state comprising the intention to accept vulnerability based upon positive expectations of the intentions or behavior of another.* (p. 395)

The term vulnerability is used here to mean that, if the other person takes an action that does not meet one's expectations, there is a risk of some damage or loss. Trust means confirming everything in advance by yourself, and in situations in which there is no certainty of safety, to take the risk and have an expectation of the other party's actions.

When one shops at a certain store, one does not suspect that the products there are fake, stolen, or damaged because they trust the store. The reason that they are able to trust a store or organization is because it has provided satisfactory services to a significant number of customers over a long period. Brand power based on history allows consumers to trust the store and conduct transactions there. If there is no trust, the consumer will have to closely trace the history and the quality of the product themselves, and it is possible that a transaction will not occur.

Research by Professor Davidson and colleagues at RMIT University in Australia estimated that as much as 35% of people employed in the US alone work in occupations that require trust.[2] There is a significant cost to providing trust to ensure smooth economic activities. Meanwhile, there is an endless number of scandals in large corporations and authoritative organizations. Rachel Botsman, author of *Who Can You Trust*? (PublicAffairs, 2017), discusses the decentralization of trust and points out that behind this is the breakdown of trust in large corporations and authoritative organizations.

According to a survey conducted by Edelman, a marketing company, trustworthiness in Japanese organizations (government, companies, media, and NGOs) is among the lowest of 28 countries, placing second to last (Figure 6-1). In addition, the trustworthiness of management is low and, in this area, Japan has placed last.[3] Trust in organizations and authorities is declining worldwide, but this decline is even more conspicuous in Japan. From the perspective of service providers, trust on servicers by consumers significantly affect the acceptance of the services they provide.

---

[2] Davidson, S., M. Novak, and J. Potts (2018). The cost of trust: A pilot study. https://papers.ssrn.com/sol3/papers.cfm?abstract_id=3218761.

[3] Edelman (2018). 2018 Edelman Trust Barometer. https://www.edelman.com/trust-barometer.

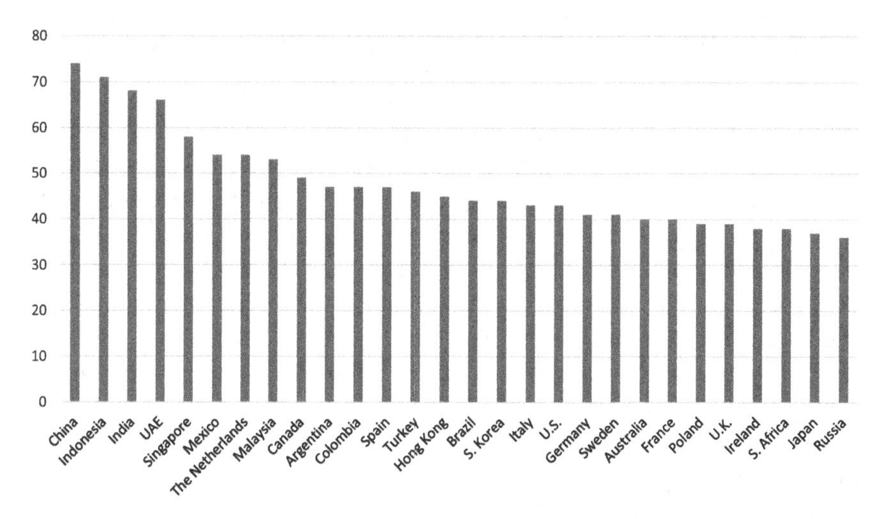

Figure 6-1 Trust level in countries
Source: Edelman (2018).

A research group of Professor Pérez-Morote of University of Castilla-La Mancha revealed that citizens' trust on government significantly influence the use of e-government services.[4]

Under these circumstances, the situation surrounding trust has changed significantly in the deframing age. The biggest reason for this is the third element of deframing, "individualization." As mentioned earlier, it is difficult for an individual to spend as much time or advertising expenses as a company would to build a brand.

Furthermore, if business is conducted over the Internet, then it is not possible to see the other person's face or to speak to them in person. Hence, in the situation where it is not possible to rely on one's trust in an organization, and where there is no way to directly confirm the counterpart in the Internet-based transactions, a major topic in the age of digitalization is how to build trust in transactions. The increasing presence of rapidly growing platform companies, such as Apple, Alibaba, and Mercari, has been driven by the companies' earnest efforts to solve the issue of trust.

[4]Pérez-Morote, R., C. Pontones-Rosa, and M. Núñez-Chicharro (2020). The effects of e-government evaluation, trust and the digital divide in the levels of e-government use in European countries. *Technological Forecasting & Social Change* 154.

Three Ways to Enable Digitalized Trust

Then, what methods do platform companies use to resolve the issue of trust? The first method is a payment escrow service (Figure 6-2). This is a mechanism in which money is transferred only after it has been confirmed that the other party has taken the expected action. This can be seen on Alibaba and Mercari: when there are Consumer-to-Consumer (C2C) transactions, the seller and buyer first agree on the transaction, and the buyer deposits the payment on the platform. The seller then ships the product, and once the buyer confirms the contents and is satisfied, the platform transfers the payment to the seller. Without this mechanism that creates a matchmaking situation, the seller would not be able to ship the product without knowing whether the buyer would actually make the payment, and the buyer would not want to make payment without knowing whether the buyer would ship the product.

By the way, there was once an online black market in the US called "Silk Road" that used Bitcoin, which was discovered by the police and shut down. However, the "Silk Road" is said to have offered a payment escrow service.[5] This illustrates the importance of how trust is mediated between strangers, even on the black market.

The second method is trust based on evaluations and reviews (Figure 6-3). As discussed in Chapter 2, it has become easier for individuals to gain knowledge on a variety of topics from the Internet and to disseminate it by themselves. In some shopping situations, information related to the experience of consumers who have actually used the

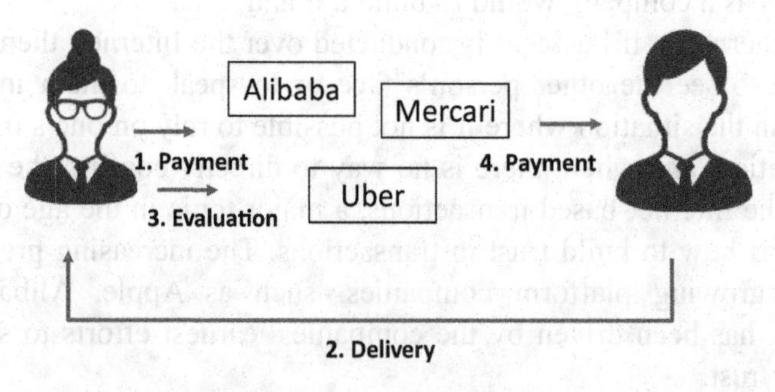

Figure 6-2 Payment escrow

[5]Narayanan, A., J. Bonneau, E. Felten, A. Miller, and S. Goldfeder (2016). *Bitcoin and Cryptocurrency Technologies: A Comprehensive Introduction*, Princeton University Press, New Jersey.

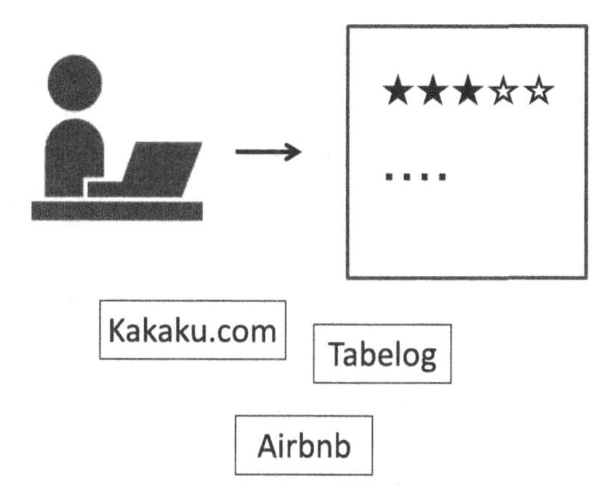

Figure 6-3 Review by users

product is more reliable than the opinion of the store clerk. In addition, consumers often know more about products than clerks. The empowerment of individuals through this "democratization of information" is the driving force that enables C2C transactions, i.e. transactions between individuals, often mediated by platforms particularly with the idea of the sharing economy.

Especially in C2C transactions, consumers who have conducted transactions in the past know best whether the other party can be trusted, or in other words, whether they can expect the product sold to them to be of good quality and whether the product will be shipped properly. Individuals who have this important information evaluate the trustworthiness of the other party in the transaction in a manner that is visible to others. This trustworthiness is usually expressed by a symbol that can be seen at a glance, such as the number of stars or a numerical value. These payment escrows and evaluations/reviews have made C2C trade possible in situations in which the other party is a stranger and it is not possible to rely on the trust of the organization.

Payment escrows and evaluations/reviews have been typical methods of ensuring trust on platforms until now, but a third method has recently emerged: the "credit economy." A typical example of this is Zhima Credit, that is provided by Ant Financial, a group company of Alibaba. This system comprehensively analyzes the creditworthiness of an individual and expresses it as a simplified numerical value. Those who already have a high Zhima Credit score are offered free services and receive preferential treatment, such as exemptions from paying hotel deposits and use of the priority lane in airports.[6] There is also a risk that the presence of the credit score can make users expose their behavior and also control peoples' act in the real world. Takeshi Yamaya, an expert on Chinese IT services reported that people try to reveal how to raise Zhima Credit score by various measures such as using credit card, not delaying repayment, make donations, etc.[7]

These credit score services have also been introduced in Japan. For example, J.Score, established jointly by Mizuho Bank and Softbank, utilized artificial intelligence (AI) to calculate users' credit score. It uses not only traditional information on occupations and incomes but also various types of information such as personal preferences, hobbies, and character, provided by the user. J.Score states these additional information enables more precise estimation of credit score, thus helps lending to the users.[8]

On the other hand, there is also a concern on the privacy regarding credit score. Yahoo! Japan used to provide Yahoo! score service, that is basically credit scoring based on the usage of the services of Yahoo! However, it was criticized by users as it seemed to provide credit information to outside companies while the score is not disclosed to the user themselves. After the criticism, Yahoo! Japan had no choice but to terminate the service.[9]

Limitation of E-Word-of-Mouth

Platforms have been working to solve the problem of trust on the Internet in various forms, such as escrow, electronic word-of-mouth (E-WOM), and credit ratings, but many challenges are still present. Among them, the typical question is whether E-WOM and reviews can be trusted. A study by Jieun Lee and Ilyoo B. Hong of Chung-Ang University showed that the adoption of reviews written by anonymous reviewers on the Internet is determined by two components: trust in a specific reviewer and review helpfulness. They empirically proved that trust in a specific reviewer is significantly affected by trust in general reviewers in the community,

---

[6] ZDNet. https://japan.zdnet.com/article/35074894/.

[7] *Ibid.*

[8] J. Score. https://www.jscore.co.jp/.

[9] NIKKEI. https://xtech.nikkei.com/atcl/nxt/news/18/08241/.

which is also affected by trust in the review website itself.[10] It shows the importance of the trust of websites and firms that operate them, which mediate reviews and comments to facilitate transactions.

On the other hand, Shinichi Yamaguchi, an Associate Professor at GLOCOM of International University of Japan, has conducted a research on this topic.[11] In a survey of about 20,000 people, 46% of the total group had posted reviews at least once. This number alone suggests that a certain number of people have experience posting something, but in reality, it turns out that a limited number of people write a large number of reviews. The people who write the most write up to 1,500 reviews per month.

Additionally, this research shows that reviews tend to contain many extreme and negative opinions that deviate from the average impressions of actual users and are not the impressions of people who have been users over a medium- to long-term period. Instead, reviews are often written immediately after use. In the distribution of opinions on the Internet, opinions are rarely moderate. There is a strong tendency for extreme opinions, such as strong approval or strong opposition, so one should be cautious to interpret the reviews. There is also a possibility that reviews are written by people who are not users. In 2012, it was discovered that there was a company that would post favorable reviews of restaurants in exchange for fees on the site Tabelog, an E-WOM platform for restaurants. Cases of bad reviews in rival stores have also been reported internationally.

Influencer marketing is another case that should be tread cautiously. As mentioned previously, when a celebrity or an ordinary person posts their impression of a product, such as cosmetics, the seller of the product may actually be paying the person. Consumers tend to believe that influencers post about products because they truly like them, but in reality, influencers may simply create the post in exchange for money. This method of relying on reviews and individuals' E-WOM is currently facing challenges because the method of confirming the authenticity and integrity of the posts is far from perfect.

---

[10] Lee, J. and I. B. Hong (2019). Consumer's electronic word-of-mouth adoption: The trust transfer perspective. *International Journal of Electronic Commerce* 23(4): 595–627. doi:10.1080/10864415.2019.1655207.

[11] Yamaguchi, S. (2018). *Economics of Flaming and Word-of-Mouth*, Asahi-shinbun shuppan (original in Japanese, translated by the author).

Blockchain as the Internet of Trust

As a result of the growth in technology in recent years, there is another important element in the decentralization of trust — blockchain technology. In my book *Blockchain Economics: A New Economic Form Through Decentralization and Automation* (Shoeisha, 2017, written in Japanese), I discussed the details and applicability of blockchain technology, as well as the possibilities and challenges of the Decentralized Autonomous Organization (DAO) with blockchain technology. Blockchain technology is so closely related to the concept of trust that it is called the "Internet of Trust." We can examine how blockchain is changing the form of trust, as well as its possibilities and limitations.

Since its creation, blockchain has been deeply involved in issues related to "organizations" and "trust." Satoshi Nakamoto, the pseudonymous creator of Bitcoin, wrote in the paper for creating Bitcoin that it was designed as a system that allows direct remittance between individuals without relying on a trusted third party such as a bank. Don Tapscott, who is known for his book *Wikinomics*, and Alex Tapscott suggested that the biggest novelty is that the transactions are certified as a result of the collaboration of many people and not trust in a large corporation.[12]

In essence, blockchain technology can be defined as "a mechanism that ensures the trust of information without relying on a specific entity." In addition, if smart contracts, i.e. software codes located and executed on the blockchain network, are used, not only static information but also dynamic programs can be shared in a manner that cannot be tampered with. Not being bound to a specific entity also raises questions about the need for entities and organizations for trust, which have been prerequisites of business up to now, which is why blockchain is said to be a disruptive technology.

Returning to Bitcoin, it carries the characteristic of money that is managed by its users, and is not controlled by any one organization or entity. Felix Martin illustrated the essence of money in his book *Money: The Unauthorised Biography*, that money is a "transferable credit."[13] There is a record of any contribution of someone doing something for someone else, and that record can be used for payment to a third party.

---

[12]Tapscott, D. and A. Tapscott (2016). *Blockchain Revolution: How the Technology Behind Bitcoin is Changing Money, Business, and the World*, Portfolio Penguin.

[13]Martin, F. (2014). *Money: The Unauthorised Biography*, Vintage.

The important thing here is that the "record" of doing something cannot be tampered with or erased or spent multiple times. Therefore, the essence of money boils down to the reliability of information.

What is the Trust Provided by the Blockchain?

If blockchain is called as the "Internet of trust," to what extent can it solve the issue of "trust" when a transaction is made between strangers? There are various levels and ways of thinking about "trust," but here I would like to consider it on three levels.[14]

The first level of trust is the question of whether the intention to trade is properly communicated to the other party (Figure 6-4). Whether the information on the intention to purchase, product information, or price is properly communicated to the other party is the most basic but important point in a transaction. Historically, this began as a verbal, face-to-face confirmation, and the restriction on distance gradually has been lowered through leaflets, such as flyers and letters. The Internet has made it possible to virtually communicate these intentions around the world.

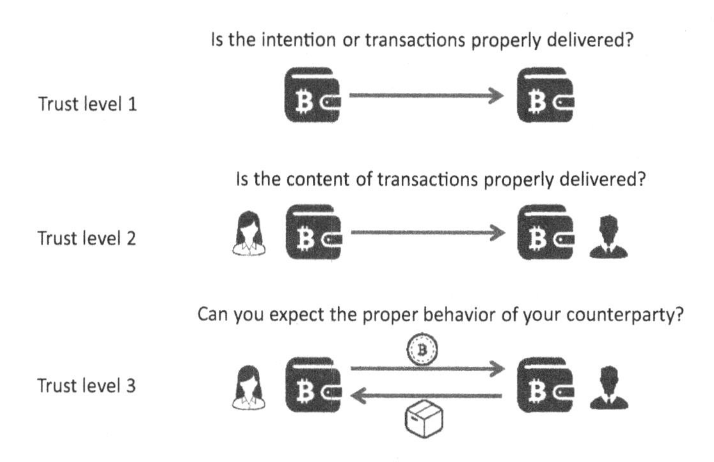

Figure 6-4 Blockchain and the concept of trust in transactions

[14]This section is based on Takagi, S. (2017). "Blockchain and Organization: From a Viewpoint of De-organization of Trust," in *Frontier of Blockchain*, Chijo Vol. 121, S. Takagi (ed.), Center for Global Communications, International University of Japan (original in Japanese).

Since the advent of the blockchain, it is thought that when the intention to conduct a transaction is registered on the blockchain, the effect is that it cannot be denied or tampered with, in the simple case of information transmission in the above. This may make some sense, particularly if the other party in this first level. is in a remote location and it is not possible to clearly confirm the party's intention to take part in the transaction. However, it is not necessary to register something on blockchain and it is sufficient in a practical sense for a user who has undergone systematic authentication using an ID and password to indicate their intention, so the simple use of blockchain may not be a significant solution to the first-level problem.

The second level of trust is whether the value is securely transmitted to counterpart. The question is whether the product will arrive and whether the currency as consideration for the product will reach the other party's account. Blockchain is new from the point that the act of sending money to another party can be reliably done without the trust of banks and remittance operators. In addition, once it is written in the blockchain, information cannot be erased or tampered with. Previously, banks took on the responsibility of managing account balances, but now this can be reliably recorded on blockchains managed by an unspecified number of people. While blockchain can successfully realize these functions to securely transmit a value, it should be noted that remittance is sending money to an address and there is no guarantee that the other party can actually receive it. If it is sent to the wrong address, and if the recipients lose their private keys, or, as is often the case these days, if the keys and the money are stolen from an exchange, it will not arrive correctly. In addition, it is also important to note that there is no guarantee that physical items such as merchandise will arrive.

The third level of trust is whether the other party in the transaction will take the expected action. The question is whether the other party will do what you expect after the contract has been agreed upon, such as whether the retailer actually ships a product after you pay for it. As we have seen, in the traditional third-party mediated model, the integrity of transaction is managed through escrow services and platform operators' guidance to retailers in addition to rewards and penalties. User reviews also act as a deterrent to prevent opportunistic behaviors. At present, it is difficult to achieve this level of trust on the blockchain, but if the smart contract program that runs on the blockchain matures in the future, it will allow for processing in which the other party's actions will be a condition of payment. Blockchain may reduce some of the risks associated with the actions of those involved in the transaction, without the intervention of a third party.

As a blockchain is referred to as "the Internet of trust," it is expected to be a mechanism that transforms and upgrade the issue around trust. However, as discussed here, actual economic transactions require broader trust and at this time, blockchain is not able to cover the entire range of these needs. As seen before, trust is "the acceptance of vulnerability based on positive expectations about the actions of others." It should be noted that there is still a long way to go before it is possible to successfully deal with people's actions and expectations by blockchain.

Incentivization Through a Token Economy

On the other hand, relating to the third level of trust discussed in the previous section, blockchain can be used to incentivize people though in indirect way. As mentioned earlier, money is a transferable debt. In other words, it is a record of some value and contribution. By utilizing digital tokens as a symbol of contributions, there is a way of utilizing blockchain to encourage people to take certain actions. Incentivizing people through tokens and building an economic space using tokens is referred to as the "token economy."

Actually, Bitcoin itself is operated by a mechanism that provides incentives to people through the token economy in the first place. The incentive in this case is the honest management of the Bitcoin ledger. Specifically, in the process of mining Bitcoin, the person who is able to finish the process called "Proof of Work" the fastest has the right to disseminate new blocks around the world. When the new block is successfully accepted by other miners, the creator of the block will receive Bitcoin as a reward for the work. The reward is halved in almost every four years and current remuneration for one block is 6.25 BTC, that worth US$246,437.[15]

Producing a new block also includes tasks such as checking double payments and the format of the data. The new block that is produced is evaluated by other miners, and if there is a defect, the block will not be used, and the reward will not be valid. In this way, Bitcoin encourages the honest actions of an unspecified number of participants by issuing tokens.

---

[15]Calculated as 1 BTC = US$39,430, as of January 15, 2021.

This improvement in trustworthiness using the token economy can also be seen in the efforts of a company called ALIS.[16] ALIS advocates "a brand-new kind of social media that clarifies which articles and people you can trust," and the key is the ALIS token. In a world where fake news is widespread, this mechanism issues tokens to people who write trustworthy articles and those who introduce those articles. While there is an issue of whether it is possible to permanently guarantee the value of a token, this initiative makes the activities and contributions to increase the trustworthiness of information visible in the form of tokens, and aims to increase the incentives for such behavior.

Trust in the token economy means that some value is reliably recorded using the trustworthiness of information based on the blockchain. Distributing it in the form of tokens encourages the honest actions of an unspecified number of people, without relying on an employment relationship.

Does the Blockchain Eliminate Hierarchical Organizations?

If we can expect people to behave in good faith without an employment relationship through the use of digital tokens, then will hierarchical organizations like traditional companies become obsolete? One school of thought asserts that, by applying blockchain, it may be possible to provide services in a wide range of fields without the need for a hierarchical organization. The concept of decentralized autonomous organizations (DAO) was born out of this concept of disbanding organizations. There have been attempts to realize various services with an autonomous and decentralized architecture, such as seen in Colony, a decentralized crowdsourcing platform; Arcade City, a decentralized ride-sharing platform; and Open Bazaar, a decentralized marketplace.

While these concepts of DAO are transforming and disruptive, it still has many challenges. Bitcoin does in fact divide the work among an unspecified number of people called "miners" because all of the work to be done can be written in a computer program, or "code." However, from the perspective of the overall ecosystem that operates payment systems between individuals, the part that is decentralized by coding is limited to

[16] ALIS. https://alismedia.jp/.

ledger management and money supply. On the other hand, there are many tasks that cannot be coded due to their high level of uncertainty. Software development of Bitcoin Core is one of these tasks, and the development community is relatively a hierarchical organization. Instead of a company, it is a community; however, proposals for functional improvements must be approved by a group of "maintainers," or they will not come to fruition. There are criticisms about the hierarchical structure of these development sites, but the Bitcoin Core community itself acknowledges on its website that such a hierarchy is also necessary for efficient development.

Additionally, decentralization in Bitcoin is actually an environment of free competition based on computing resources. As mentioned previously, it is the task of the "miner" to update the ledger, but which miner will conduct the task and then receive the newly issued Bitcoin as a reward is decided by the competitive process of "Proof of Work."

The process is an environment of free competition, so winning depends on having a high-performing computer and being able to cheaply procure the electric power to run it. As mentioned earlier, a semiconductor chip called ASIC is used specifically for Bitcoin mining. Without a "mining factory" in which there are a large number of ASIC-based computers, it is extremely difficult to win the mining competition. Initially, Bitcoin was supposed to be a currency managed by the users themselves, but it is no longer an environment in which an amateur can enter and find success.

In a decentralized world, it is necessary to decide how to share works among an unspecified number of people, but to share the work fairly and with incentives, it is convenient to make decisions based on the competitiveness of simple tasks that anyone can undertake. This is how Bitcoin was created, but ironically, the result is that mining has become an oligopoly. Furthermore, in examining the payment ecosystem that uses cryptocurrency, not only blockchain but also various centralized organizations such as cryptocurrency exchanges, digital wallet operators, and Initial Coin Offering (ICO) issuers are connected. These centralized organizations also play an important role in the Bitcoin ecosystem.

In addition, the risks associated with managing private keys have become apparent in repeated cryptocurrency outflow incidents. Initially, the idea behind decentralized Bitcoin was that the private key would be strictly managed by the user, but general users do not want to be bothered with this. Users want a professional who can securely manage their private keys, and expect the organization to take responsibility in the event of a problem. The outflow incidents have also highlighted how deeply rooted consumer expectations are toward a centralized organization.

While Bitcoin has worked in earnest for "decentralization," QR-code–based payments such as Alipay and WeChat Pay in China are highly centralized payment methods. Japan is also experiencing a similar boom of payment apps. This widespread boom shows that if convenience and benefits far outweigh the concerns on centralized operators, people will use a technology. Cryptocurrencies based on blockchain technology such as Bitcoin have been developed with an emphasis on privacy. However, with the spread of centralized "cashless" payment methods and depositing private keys of Bitcoin to exchanges, it may be necessary to reconsider how people weigh between privacy and convenience.

Blockchain was created with the purpose of eliminating the need to trust an organization, but for now, it seems difficult to eliminate everything. Currently, it is difficult for blockchain to completely eliminate hierarchical organizations. There is also the issue of how much consumers value decentralization.

Towards a Trusted Decentralized Trust

In the deframing era, individuals will play a more important role in industries, bringing together more detailed business elements. Therefore, in economic transactions, we cannot rely on the lubricant of brand power that large companies have built up over the years. Finding the complement to trust for decentralized and distant economic entities is an important factor in the future economy, and if it can be resolved, it becomes a huge business opportunity.

As discussed in this chapter, several ways to complement trust have emerged. One is escrow payments through a platform, which is a mechanism that encourages both parties to be honest in conducting transactions. Reviews can also reduce uncertainty about service and product quality. Additionally, although the blockchain is operated by an unspecified number of people, cryptography can be used to ensure that the recorded value will not be lost or stolen.

However, these methods still pose many challenges. Although mediation using a platform is convenient, this method relies on a particular company for a key part of the economy, which can lead to an extremely uneven distribution of wealth and privacy issues. Manipulation of reviews and the existence of stealth marketing are also issues. Complete decentralization has not been achieved by blockchain technologies.

Ultimately, trust fills the gaps where there are human limits to the capacity to process information. As discussed earlier, trust is about accepting risks in anticipation of others' actions. Trust is not necessary if you are able to discern all the risks in a transaction and take steps to avoid them. In the case of online shopping, if it is possible to thoroughly examine all of the information about a store and its products, in addition to taking measures in case of an unexpected problem, issues related to trust may not arise. However, consumers generally do not have enough time to spend doing this. When an organization can offer trust based on many years of conducting transactions, the consumer does not have to bother with the task of verification. Whether it is better to trust an organization or to have decentralized trust depends on the consumer.

On the other hand, when trust is decentralized, it is possible to receive trust that is more optimized to the individual. It goes without saying that more individual and specific evaluations are more useful, such as the quality and usability of a specific product, or the impression of people who have been in the same situation as oneself. It is important to know how to improve the reliability of such personal information and how to complement an individual's limits in their capacity to process information. As deframing progresses, there will be a greater demand for systems to ensure trust in more specific perspectives. This could uncover hidden business opportunities for companies.

Chapter 7

Personal Strategies in the Age of Deframing

Thus far, we have discussed the challenge of dealing with the decentralization of trust, after looking at the three elements of deframing, namely "dissolution and reintegration," "specific-optimization," and "individualization." What kind of career should we build in this era? In this chapter, we will consider personal strategies for surviving the era of deframing. To that end, I would like to first provide an overview of the major social trends surrounding individuals, then discuss the human resources and workstyles that will be required in the future, and what we can do to achieve this. Please note that social systems and conditions such as employment, social security, and pension plan are different across countries. The arguments in this chapter are mainly based on those in Japan, that are characterized by super-aging, mandatory retirement system, and public pension plan that is pay-as-you-go. However, many of the implications would apply for many countries that share the same problem such as aging.

Super-Aging Society and Prolonging Career Time

The issue of declining birthrate and aging population has had a major impact on our workstyles. In every country, life expectancy increases as the economy becomes wealthier, while the birthrate is in decline due to the impact of increased education costs and later marriages. It is well known that the issues of the declining birthrate and aging population are particularly significant in Japan, but super-aging has a great impact on workstyles and career design due to issues caused by raising the pension age and that accompany employment in the older age.[1]

From the perspective of an individual's career plan, the first impact of a super-aging society is on social security. While in 1965 in Japan, the "*dōage* model" (many people supporting one person) was in place, as there were nine people of working-age supporting one elderly person, but by 2050, the "*kataguruma* model" (piggyback ride) will be in place, when there will be 1.2 working-age people supporting one elderly person.[2] The current social security system of Japan, which is based on a pay-as-you-go pensions in which the working generation supports the elderly, rather than accumulating savings for retirement, is the area most affected by aging. In this scenario, the burden on the working generation may exceed sustainable levels.

Conversely, it cannot be said that issues related to retirement funds are solely due to pay-as-you-go social security. Figure 7-1 is a conceptual

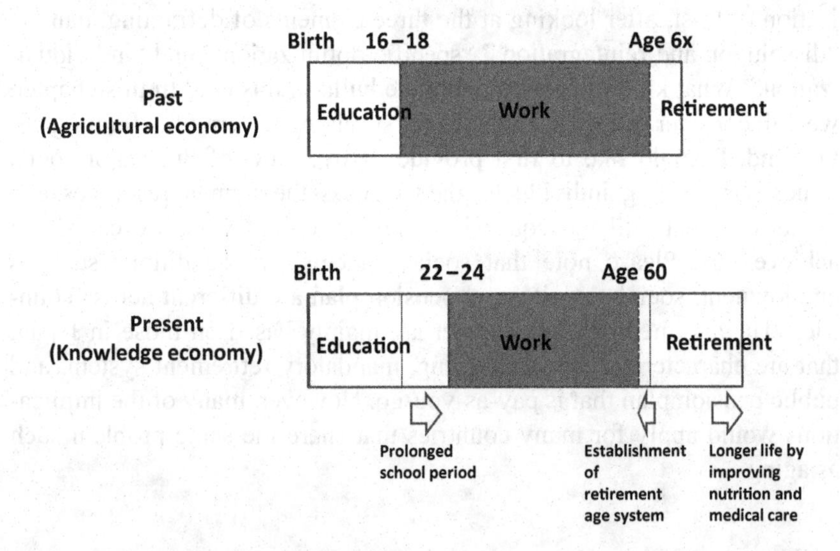

Figure 7-1 Comparison of working time through life

[1]This section is partly based on Takagi, S. (2016). Considering career flexibility suitable for the super-aging era. *GLOCOM OPINION PAPER*, No. 5 (original in Japanese). http://www.glocom.ac.jp/wp-content/uploads/2016/10/OpinionPaper2016_No5.pdf.

[2]Ministry of Finance. Overview of integrated social security and tax reform (original in Japanese).

comparison of working periods in the past (agriculture-based society) and the present (company-based society).

As mentioned in Chapter 2, in the era of the agriculture-based society, production activities begin immediately after one graduates from junior high or high school. Since there was no system of mandatory retirement, people worked while they were healthy and then gradually retired. The average lifespan was shorter, so the period of "retirement" was also short, and it is inferred that at least about 60% of one's life was spent working. In modern society, the years of schooling are increasing, and we start working at a later age due to the knowledge economy. It is now common to get a job after graduating from a four-year university or two-year graduate school; if the knowledge economy progresses further, then it is possible that there will be more jobs that require a postgraduate or master's degree, or a doctoral degree.

As a result of the shift from a family-owned, agricultural society to an industrial society centered on office workers, the mandatory retirement system has become widespread, and it has become common to leave the labor market around the age of 60. Therefore, it is estimated that the working period in life has been shortened around 40% or 50% due to the dramatic increase in lifespan. Even if this is considered from an individual perspective, it means that a person needs to earn double their living expenses, otherwise the figures do not add up. Even if the pension system is an accumulated saving system, it will not be easy for individuals to balance their income and expenditures over the course of their lives.

There are some views that older people are healthier and behave as though they are younger than the elderly of the past.[3] While they still have the ability and physical strength to work, there are situations in which they are forced to be "supported" by the systems such as mandatory retirement. According to a survey jointly conducted by GLOCOM, International University of Japan, and Panasonic Corporation Media & Spatial Design Laboratory, it was found that about 75% of people want to work after the age of 60, while only about 35% of people actually work after the age of 60.[4]

---

[3]Akiyama, H. (2012). *Longevity Society and ICT* (original in Japanese, translated by the author). http://www.soumu.go.jp/main_content/000190764.pdf.

[4]Questionnaire survey of 6,000 Internet monitors. Conducted in March 2016.

The Illusion and Challenges of "Lifetime Employment"

Under these circumstances, there are many initiatives to encourage the elderly to work. A typical example is the extension of the retirement age and re-employment of retirees, however, there are many problems that accompany these initiatives. The survey mentioned above shows that the majority of people who have not worked after the age of 60 were previously full-time employees of a company. Among them, the ones that particularly face problems are surprisingly those with experience in managerial positions. Those who were the head of their section or department tend not to, or cannot, work after the age of 60. Once a person becomes a manager, they may withdraw from the labor union, and because of the mandatory retirement age for managerial positions, they may be seconded to a subsidiary at an early stage. It is unique system of *Yakushoku Teinen*, seen in Japanese large corporations, where managers must quit the position at the age of around 55 and transferred to subsidiary companies unless they become corporate executives. Thus, even if they possess the ability to work, they lose the chance to demonstrate their capability, and are excluded from requirement to continuously hire those in their old age.

In addition, there are some issues related to the standardized extension of retirement age and the re-employment of retirees. There are issues regarding whether the retiring employees are suitable or unsuitable for a fast-changing business, and organizational management issues may occur when one who was a manager becomes a subordinate. Also, in today's rapidly changing world, there is the issue of whether a company will be stable and remain in business through employees' long lifetimes. Even if the mandatory retirement age is extended, there is a possibility that the company itself may not last that long.

The idea underlying the term "lifetime employment" is that the employee will serve the company over one's lifetime and in exchange do not have to worry about income throughout their professional life. However, as mentioned previously, in this era of longevity, even there is a "mandatory retirement age," it can no longer be referred to as "lifetime employment." From the perspective of a company, guaranteeing employment for all employees up to the age of 65 or 70 will not be easy, considering the competitiveness of the industry and the impacts on new hires. Given these issues in a super-aging society, there is a high risk that workers will entrust their lives to one company.

However, the skills and knowledge that a person cultivates over many years are valuable assets not only for their company but for society as a whole. The main principle of deframing is to make effective use of valuable resources. To make effective use of valuable human resources, companies need to seriously consider flexibility in forms of employment. For example, many Japanese companies have severely restricted their employees from holding second jobs. However, assuming that the company will not be able to take care of its employees by employing them for life, they should indeed be flexible and allow second jobs.

If a person can shift their main job in the course of their age, by working for one main company while holding different jobs and income sources, this will become a completely new safety net in addition to their pension. It is also possible for a person to start a business in their senior years or be hired by a venture company, based on the experience cultivated at their main employer. Making use of that experience is not only beneficial to that person, it is of great significance to society as a whole, which suffers from human resource shortages and a stagnation of innovation.

In response to these changes in the environment, the Japanese government announced the action plan for work style reform in March 2017 and decided to formulate guidelines toward allowing side jobs and multiple jobs. After that, the Ministry of Health, Labor and Welfare announced a model of the guideline for promoting side jobs and multiple jobs, and the model rules of employment including provisions for side and multiple jobs.

The COVID-19 pandemic further stimulated the actual conduct of side jobs and multiple jobs. Nikkei reported that one million "gig workers" newly emerged only in the first half of 2020 in Japan.[5] On the other hand, Mizuho Financial Group, one of the largest financial groups worldwide, took a bold step to allow its employees to work side jobs, aiming at increasing capability of employees with various business experiences.[6]

In this modern age of super-aging, it is important to design your own portfolio of skills and professions so that you can protect yourself, without being bound to the illusion of "lifetime employment."

---

[5]NIKKEI. https://www.nikkei.com/article/DGXMZO60710750T20C20A6EA2000/.

[6]NIKKEI. https://www.nikkei.com/article/DGXZQOGH1912N0Z11C20A2000000/.

Can Basic Income Become the Key to Solving Technological Unemployment?

In recent years, there has been a great deal of debate about whether the rapid development of artificial intelligence (AI) may deprive humans of employment. A report published by Oxford University researchers sparked these debates, stating that half of jobs could be lost to AI.[7]

In fact, the decline in employment that accompanies these technological innovations is a phenomenon that has long been known as "technological unemployment." Looking back on human history thus far, the invention of the automobile certainly eliminated the work of the human-powered litter and coachmen who used horses. In addition, the development of tractors and fertilizers has made it possible to manage crops with a small number of people, dramatically reducing the number of farmers. Furthermore, automation in factories has made it possible to produce products with significantly fewer people. Economics textbooks state that these eliminated jobs will be replaced by more productive professions. Certainly, throughout our history, employment has been created by such logic. However, the problem with today's AI is that if it continues to evolve at such rapid speeds, shifts in occupations will not be able to keep pace, creating a large-scale unemployment.

If occupations disappear over a few generations, then children and grandchildren must be educated differently and are able to engage in different occupations. However, if technology spreads at a speed that is so quick that people have to change jobs while they are working, this change will be accompanied by a great burden on workers. In addition, some argue that computers will someday eliminate the need for any occupation.

In response to these concerns, there is a movement that proposes a universal basic income as the solution. Universal basic income means that the government unconditionally pays everyone a certain amount of living expenses. The idea is to eliminate any current public assistance and pensions and provide a uniform, fixed amount (e.g. 70,000 JPY per month in living expenses) to both those who earn money and those who do not. This way of thinking has been discussed from various perspectives such as

---

[7]Frey, C. B. and M. A. Osborne (2013). The future of employment: How susceptible are jobs to computerisation? https://www.oxfordmartin.ox.ac.uk/downloads/academic/The_Future_of_Employment.pdf?link=mktw.

human rights, compensation for domestic labor, and for the eradication of poverty. It has attracted attention, based on the idea that if most occupations disappear due to automation, basic income should be introduced.

Basic income has the advantages of providing livelihood support for all and reducing the cost of public assistance screening and paperwork. In addition, non-wage labor such as full-time housekeepings will also be able to earn income; because their income is guaranteed, it is expected that it will become easier for people to take on challenges. Conversely, there are concerns about whether financial resources can be secured and whether the motivation to work will decline.

This model was recently rejected in a national referendum in Switzerland. In addition, the model was introduced in Finland as a pilot program, but was canceled at the end of 2017 without reaching full implementation. A similar pilot program was conducted in Ontario, Canada, but the discontinuation of the program was announced in July 2018.

Basic income itself has several advantages, as mentioned above, and is a concept that deserves consideration, but it is also difficult to use as a measure against unemployment that results from the spread of AI. It is well known that services using the Internet are easily provided across national borders. Whether it is a search service or shopping, people all over the world can use the most convenient service. This advantage is strengthened by the network effect, in that the more people use it, the more convenient it becomes.

The platform services discussed in this book in particular have strong network effects, and the more they are used, the more data is accumulated and can be used for "learning" in AI, so there is an increased possibility that large platforms may be able to develop better AI. However, even if the most convenient AI developed in this way "earns" enough to replace human occupations, the AI may be one that is operated by overseas business.

If AI replaces employment and the lack of income is compensated with basic income, it will be necessary to tax services that is provided with overseas AI. Of course, it would be necessary to assume that the opposite position would also be true. For example, foreign governments may seek to distribute the added value created by AI in your country. Today's services are globalized, and a country's social welfare system can no longer ignore the flow of wealth on a global scale. Considering this from the principle of deframing, it may be impossible to consider social security within the framework of a country. If AI will eliminate employment and the solution for this is basic income, then there will be a need to integrate social security mechanisms globally.

In addition, what should we think about the idea that if AI eliminates the need to work, we can just enjoy our lives? This certainly may be ideal if one does not mind possessing very basic clothing, food, and housing, comparable to the standard of living in the Medieval period. Food production, garment production, and housing construction may be produced mostly automatically, even at current technological levels. However, human beings are creatures that cannot be satisfied. As long as a person has the desire for something a little more comfortable, something unique to themselves, and something they can feel satisfied with, there will be a need to work to add that "something more" to their current situation. As long as we are not satisfied with the bare minimum, we will continue to use our human ingenuity, or in other words, our work.

When AI is the property of a group of people or a company, the concept of basic income for technological unemployment caused by AI means a system in which the earnings generated by AI should be distributed. In an extreme sense, rather than pledging allegiance to the AI that generates earnings, people receive a distribution as a "salary." Put another way, there is a fine line between basic income and digital feudalism. After all, the capital strength of the company that owns the AI is used to provide convenient services. Even if individuals use AI in their work, each person needs to learn how to use AI effectively, and there is a possibility that work will be concentrated to the people who can use it even a little better than others.

One book that has become a popular topic of discussion is *Homo Deus* by Yuval Noah Harari, in which the keyword "the useless class" appears. The argument goes, as a result of AI and machines being able to do everything better than humans, humans will be unable to make any social contributions. The inability of humans to make social contributions in this way is a threat to liberalism, which is based on individual dignity. The idea that people can live off basic income to enjoy a life of leisure if work is eliminated due to AI may, in fact, threaten the foundation of liberal democracy.

Similarly, it is dangerous to think that people can use basic income to live lives of leisure. In reality, such a day may not come, and if it does, it may be the day when humans lose their independence and individuality. Even if AI provides useful functions, perhaps it is more likely that we will continue to need activities to make full use of the AI to provide new services and to solve problems on our own.

Career Strategy in the Age of Deframing

As we have seen, in the era of deframing, the traditional "frame" has less significance. As a result, the importance of belonging to a company, a workplace, or even university, and over the long term, national and local governments, will become vague. This will have a significant impact on one's life plan, and especially on the idea of working for a company. In the past, it was common to join a company upon graduating from school and spend one's life there until mandatory retirement. In such a society, it was sufficient to meet the skills required for one's specific job title and occupation rather than focusing on the unique abilities of each individual. Instead, because it was assumed that employees would be reassigned every few years in large Japanese companies, it was more important to have the ability to handle various tasks. In addition, rather than a professional who is highly versed in the business, more value has been placed on the understanding of company-specific culture and protocol. This is likely because the role of the company in society was relatively stable, and both companies and individuals have been protected from competition within the framework of the company, so there has been leeway for this to occur.

However, the deframing era is one in which the best thing is sought out and used, whether it is inside or outside a company. The basic idea of deframing is to make good use of the precious resources that already exist, and this also applies to individual skills. In an environment in which matching and transactions are rapidly accelerated by the power of technology, even if you are a company employee, it is necessary to think of the competition with an entity outside the company.

To counter this situation, it is important for each individual to hone their unique knowledge, skills, thought patterns, areas of expertise, and interests, and utilize them without being bound by the framework of the organization, in addition to knowing how to connect with others. It is also necessary to re-learn as the environment changes.

Importance of Becoming Unique

As previously touched on, in the current world of advertising, influencer marketing is on the rise, in which ordinary people and celebrities who are influential on social networking services (SNS) are given rewards in exchange for introducing products. In general, influencer marketing is a method in which a company provides products and/or compensation to people who have the power to communicate well on SNS or blogs these people introduce the products in exchange with the rewards. In addition, even if the company does not necessarily pay money, it has become possible to run a business that earns advertising revenue as long as the person has the ability to communicate well, as YouTubers have shown. The workstyle of "influencers" has become established through their individual earning method.

Initially, one of the features of this method was that anyone could be an influencer, but as competition intensifies, the situation is changing. One reason for this is competition: because anyone can become an influencer and can disseminate information, it is necessary to continue to create high-quality content in order to have consistent viewing. If this becomes the case, it will become difficult for an amateur to continue to make content the same way, so they might join an entertainment agency and become partly professional. For example, the company UUUM offers management production services for YouTubers and also provides support for YouTubers from content production such as shooting, planning, and product sales. At this point, aiming to be a YouTuber in this way is similar to aiming to be an entertainer by approaching an entertainment agency or attending a vocational school for comedy. On the other hand, it is not clear if this trend will continue: when information on how to create better content becomes widely available, an individual would be able to independent again to be competitive in YouTube market without professional agencies' support.

Another aspect is that this area is progressing toward complete "nichification." Influencers continue to pursue themes that are specialized to specific needs, such as solo camping, travel with specific themes, cosmetics for cosplay, chefs who cook meals in a short time, and photographers who specialize in picturesque scenery. In an era in which there is an abundance of information, if a person does not show their uniqueness, they may not be discovered or maintain loyal audiences. In fact, this trend applies to all industries. These are times in which the world is connected through the Internet and anyone can search for the best of anything. At the same time, anyone can become a provider. However, those who will be found from the vast number of providers will either be the all-mighty superstar, or someone at the top of their narrow niche area.

Whether someone aims to be a "superstar" or a "niche" YouTuber, an important strategy in the future will be to aim to be a number one in the range that information is distributed. It is a tough environment to be in second or third place, so it is necessary to be more aware of differentiation than before. Without differentiation, it is possible to get caught up in price competition on a global scale. Price competition is already evident in the crowdsourcing world. A total of two million people are registered for crowdsourcing in Japan,[8] but according to a survey in 2013,[9] about 68% of non-enterprise crowd workers receive less than 5,000 JPY per month, on average. In order to be matched through a crowdsourcing platform, the content must be standardized. When it comes to tasks that many people can do, such as simple web design, translation, and logo creation, price competition ensues. The challenge is how to hone one's unique abilities in a certain field and escape from standardized work.

Thus, it becomes important to ensure that one's special abilities can be discovered among the overflow of information. In terms of being discovered, design will become increasingly important going forward. The term design has various meanings depending on the field, and the question is how to organize interfaces and relationships between various elements, which is especially deeply related to human cognition and understanding. Design is an important skill in the deframing era from the perspective of how to effectively convey one's strengths and selling points to another party.

Leadership and Re-Learning in the Deframing Age

The deframing age is an age of individualization and there may be the impression that it is unrelated to leadership. However, leadership is becoming increasingly important because, until now, leadership needed to be exercised on the premise that one had the authority as a boss in a hierarchical organization. However in the future, it will be necessary to lead a team that does not necessarily include the relationship between a boss and their subordinates. In the deframing age, even if people work together, they may actually belong to different organizations. Of course, if there is

---

[8]Ministry of Health, Labour and Welfare (2015). *Current Status of Crowdsourcing* (original in Japanese, translated by the author). http://homeworkers.mhlw.go.jp/files/h27-report-crowdsourcing.pdf.

[9]The Small and Medium Enterprise Agency (2014) 2014 White Paper on Small and Medium Enterprises in Japan.

a contract in place, it might be possible to give instructions to some extent, but if the other party does not like it, they may choose to end the relationship.

Thus, going forward, motivation management becomes important instead of command and control. Because individuals who originally worked separately under different organizations and for differing incentives are temporarily teamed up, the key factors of leadership are the ability to know the aim and significance of a project and guide members in a way they can understand. For example, Reed Hastings, CEO of Netflix and Erin Meyer, a professor at INSEAD, illustrated that the management in Netflix, a major video broadcasting company, has the characteristics of "leading with context" rather than leading with control. It focuses on sharing every contextual information and encourage autonomous decision-making. At the same time, they suggested that this management style requires high density of talented workers.[10] In other words, leading with context works well when workers are self-motivated as often seen in entrepreneurs, and in this sense, this concept is typically applicable to the organizations that took the elements of individualization.

At the same time, it is necessary to have a broad understanding of what kind of human resources exist and where they are. It is important to have a wide range of personal connections, such as someone who is familiar with the latest business model trends, someone to develop mobile apps, someone to approach a certain company, and to be able to behave as if they are a platform. Traditionally, senior-level employees in traditional companies have had this function similar to a platform only within the company. The presence of these senior-level employees was valuable in an age when work was completed entirely within a large multi-divisional company. To be a leader in this coming age, it will be necessary to build a wide range of personal connections by making full use of SNS and connecting on-demand, rather than secluding oneself within the company.

In the age of deframing, there is an emphasis on how to connect individual abilities and skills, and work is advanced by forming a team for each project. Therefore, one must have an awareness of the skills and strengths that they possess. Additionally, it is necessary to

[10]Hastings, R. and E. Meyer (2020). *No Rules Rules: Netflix and the Culture of Reinvention*, Penguin Press; WH Allen, London.

continually update one's strengths, considering that careers are being extended due to the super-aging society. In this sense, re-learning is becoming important.

According to the Edelman survey of freelancers introduced earlier,[11] 55% of freelancers have worked on improving their skills through education and training in the past six months, but only 30% of regular employees engage in this. In addition, freelancers have a stronger feeling of threat that their work will be eliminated by computers. Perhaps freelancers are more aware of the importance of re-learning.

In addition, in the survey on employment in old age introduced earlier,[12] those who were engaged in the "acquisition of qualifications," "funding for entrepreneurship," "acquisition of knowledge and technology," "work at other companies," and "acquisition of skills for a sole proprietorship," tended to have a high employment rates in their old age. Being engaged in "re-learning" while still at working age, for the purpose of a career shift, is also useful for working in old age.

On the other hand, deframing makes it easier than ever to learn new skills. With online education apps such as Udemy and Dedao, which were mentioned earlier, it is possible to learn various skills, such as the mechanisms of AI, how to create web services, web design, entrepreneurship, and self-development, at a reasonable price.

The ability to update one's skills as needed in this way is also a countermeasure against the declining birthrate. One of the reasons for the presently declining birthrate is that the knowledge economy has progressed, and education has become more expensive. When it becomes commonplace for everyone to go to college or graduate school, the financial burden to raise even one child could be severe. At the same time, the longer a person spends being educated, the later they will get married, and they may lose the opportunity to have children. If it becomes possible for someone to easily update their knowledge at any time during their career, rather than receiving all their education before they start working, it could lead to a reduction in the burden people feel in regard to education.

---

[11] Edelman Intelligence (2017). *Freelancing in America: 2017* https://www.slideshare.net/upwork/freelancing-in-america-2017.

[12] Jointly conducted by GLOCOM, International University of Japan, and Panasonic Corporation Media & Spatial Design Laboratory.

Education in the Age of Personalization

Fortunately, the educational system itself is beginning to move toward the second principle of deframing, i.e. specific-optimization. There is a movement toward providing education that matches individuals' progress, areas of strength, and intentions. Some people may have had a difficult time learning math, English, or statistics in the past. It takes a great deal of patience, especially in subjects that need to be studied systematically, as each step needs to be understood and memorized. In his book, *Learn Better*, Ulrich Boser provides a wide range of explanations about what learning is and how to learn deeply. He states that "the goal of learning is about shifting how we think about a fact or idea, and when we learn, we aim to learn a system of thought."[13] This means that rather than simply memorizing factual information, it is necessary to understand something deeply enough that you take on a new way of thinking.

Boser states that being able to recognize that the subject of learning is valuable to oneself is the first step in learning. To achieve such deep learning, it is important to know how to connect each learner's interests and concerns related to the subject being studied. This is an opportunity for teachers to showcase their skills. Knowing what kind of relationship exists between each student's interests/concerns and what he/she is going to learn, and how to make the student understand this will have great influence on the subsequent effectiveness of what was learned. There is a problem with traditional, one-way learning in packaged education. In traditional lessons in the classroom that simply transfer knowledge, it is not easy for students to connect their interests and concerns with the subject. If they are unable to find meaning in a subject, they will certainly not be interested in learning. When it comes to subjects that are even more difficult to understand, it will be difficult to continuously invest the energy to learn.

One of the possible approaches is to determine the subjects that will be studied in accordance with the student's intentions. Fortunately, there is a vast amount of content on the Internet today to help you learn what you want to learn. For example, the previously mentioned platforms in the educational field, Udemy, Schoo, and China's Dedao, allow people to

---

[13]Boser, U. (2017). *Learn Better: Mastering the Skills for Success in Life, Business, and School, or How to Become an Expert in Just About Anything*. Rodale Books, United States of America, p. xx.

learn about various topics, such as web development, illustrating, design, foreign languages, investment management, and how to start a business. If you choose what you want to study according to your own interests, you will be more motivated to learn because it is linked to your interests and concerns from the start.

In "AltSchool," leaving the discretion to the learner has been adopted in school education. AltSchool is a school founded in San Francisco in 2013 by former Google employees and provides education for pre-kindergarten through 8th grade in San Francisco and New York.[14] Here, the curriculum is designed by the student, and the student also decides their own homework and personal goals.

Personalization here is possible because educational content and homework submissions are fairly systematic, and many apps are also used. This method makes use of the second principle of deframing, specific-optimization through mass customization. Another similar initiative is "Summit Public School," which is affiliated with Facebook. Here, an IT platform called the Personalized Learning Platform (PLP) was constructed so that students can design what they will learn by themselves. In other case, Qubena, an educational service provided by COMPASS Inc, provides personalized education service by using AI. For example, it analyzes the points students found difficult in mathematics using AI and optimizes the flow of learning.[15]

On the other hand, issues have also been identified in personalized learning. There is some research that pointed out that, compared to mass learning, the progress of learning is slower.[16] However, according to a survey supported by the Bill and Melinda Gates Foundation,[17] this method of personalized learning has been seen to be particularly effective for math and reading comprehension, with a significant positive effect on low-performing students. Students who score highly across a wide range of subjects may still perform well even in a mass learning situation, but some of their abilities may not be fully developed. This is an area that requires further research in the future, but as the knowledge-based economy

---

[14] Altschool.

[15] Qubena. https://qubena.com/service/math.

[16] TechCrunch. https://jp.techcrunch.com/2017/11/24/2017-11-22-altschool-wants-to-change-how-kids-learn-but-fears-that-its-failing-students-are-surfacing/.

[17] Pane, J. F., E. D. Steiner, M. D. Baird, and L. S. Hamilton (2015). *Continued Progress: Promising Evidence on Personalized Learning*. RAND Corporation.

advances and education becomes more important, providing education that suits each individual is becoming an even more important issue.

The Ultimate Goal of Education is to Learn How to Learn

We are no longer in an age in which we can acquire knowledge and then live off of it for the rest of our lives. In today's world of extremely rapid technological progress, acquired knowledge and skills quickly become obsolete. Table 7-1 shows the transition over the past 30 years in the major programming languages, published by Dutch software company TIOBE.[18] In 1998, C, Lisp, and Ada were the top three languages. As of 2018, Java, C, and C++ are in the top three, with Python growing at a rapid pace. Furthermore, of the top 15 in 2018, nine did not exist in 1988, or if they did, they were not at the top of the list. C is still in use, but for those who learned Lisp and Ada 30 years ago when they were new employees, if

Table 7-1 Major programming languages

Programming Language	2018	2008	1998	1988
Java	1	1	17	—
C	2	2	1	1
C++	3	3	2	4
Python	4	6	24	—
C#	5	7	—	—
Visual Basic .NET	6	—	—	—
PHP	7	4	—	—
JavaScript	8	8	21	—
Ruby	9	9	—	—
R	10	48	—	—
Objective-C	14	40	—	—
Perl	16	5	3	22
Ada	29	18	12	3
Lisp	30	16	8	2
Fortran	31	21	6	15

[18]TIOBE. https://www.tiobe.com/tiobe-index/.

they want to continue their career as an engineer, they need to re-learn a new programming one.

However, differences in programming languages are sometimes referred to as similar to dialects, and once learned, conversion is generally not that difficult. When someone first learns a programming language, they will have some degree of experience with the basic principles, how to use it, where to be careful, and the learning method. By reusing that system of knowledge and skills, learning a new language will be much faster than the first one.

This is not limited to programming languages, but is common to all forms of learning. Learning a foreign language, learning accounting methods, or gaining skills for performing managerial duties are seemingly completely different to learning golf, skiing, or learning to play a musical instrument. However, these things do have common elements, and if you learn several ways of learning that suit you, this will help you learn a wide variety of things. For example, the following are some common ways of thinking that can be used in learning.

The first one is to dissolve the content and work on the smallest possible unit. If there is something you do not understand or is difficult, dissolve it into elements. If there is something in your studies you do not understand, go back to the point where you do understand, and look up only the terms and rules you do not understand. This method of dissolving and tackling the finer elements one by one will allow you to progress forward, even if it is only slightly. The second way is to get organized in your own way. If you skim through a textbook, it might not actually stick in your brain. When this happens, create a relationship diagram, or something similar, of the concepts by yourself, and make sure you are able to explain it to someone else. Explaining something in your own words will help you understand it more clearly. This method is especially useful for grasping the overall structure rather than the details. If you can give a short explanation of what it is, this will be one piece of knowledge on the subject. The third point is to set goals. It is also useful to set time-bound goals, such as setting a goal to take a qualification exam, or give a presentation in front of your colleague three months from now. If you only set the qualification test as a goal, you can postpone it at your own discretion, so it is a good idea to make a promise in front of others. These are just some examples, but if you are able to acquire your style of learning, you will be able to respond flexibly and enjoy when you encounter a situation in which you learn new skills.

On the other hand, it is a major issue whether this type of learning experience can be offered in school education. For example, in physical education, telling a student to run x kilometers can only be painful for them. In running, there are various elements of the skill, such as how to use your core, how to swing your arms, and how to balance yourself. Only after learning these skills would you feel the "fun of running." When a student feels excitement and surprise such as, "I was able to run much easier than before" or "I could run faster," can this be called education.

If you feel this excitement and surprise about learning, it will be useful when you try to acquire new knowledge and skills. Acquiring methodologies and systems for learning, and experiencing the excitement being able to do something you could not do before can be a tool and a driving force for learning new things. We are now in an age in which re-learning is continuously important. To meet the needs of this age, teaching people how to learn and about the joy of learning is extremely important.

Chapter 8

Challenges and Prospects of Deframing

Thus far, we have examined the concept of deframing and its specific examples, in addition to considering how individuals should respond to the process. Deframing is a phenomenon inevitably brought about by technological development; however, the future is not necessarily rosy. There are positive and negative aspects to any kind of change. The way in which we reduce the negative aspects of the social impact that technological progress inevitably brings is an extremely important point of view in the present age. In this final chapter, I would like to first discuss the downsides of deframing and consider the solutions, and expand upon this with prospects for the future and perspectives that have not been addressed so far.

Oligopoly of Platforms

The first issue is that of platform growth and oligopoly. The principle of deframing dissolves the individual elements that have been packaged and connects them to construct new services and products. Therefore, the economy has become "decentralized" and it seems that we are moving toward a more equal and fair society with less disparity. Certainly, if we can make good use of the talents and resources that were previously hidden, this will increase the likelihood that potential value will materialize in every corner of society, which will lead to the creation of wealth. However, this would not necessarily achieve equality in terms of wealth distribution.

One problem is that we currently need a platform that connects individual elements. Such a platform should collect, organize, analyze, and match information from various individual elements, enabling scarce resources to be utilized in society. Crowdsourcing, peer-to-peer (P2P) lending, and advertising optimized for an individual would not be possible without a platform. An essential condition of deframing is the mechanism that allows various elements to meet and be matched in one place. When dissolving and reconnecting elements that were previously packaged, a platform to allow matching must exist. The rapid progress of deframing and the rapid growth of platforms are two sides of the same coin.

Furthermore, platforms are generally more valuable the larger they are. It is possible that you can sell something better in the place where 1,000 people are participating than where only 10 people participate. With these network effects at work in the background, the platform is likely to become monopolized and oligopolistic, and as the transactions increase, the more wealth is concentrated on the platform in the form of fees. Table 8-1 shows the global market capitalization ranking of companies as of the fourth quarter of 2020.[1]

Among the top 10 companies, 7 are structured with an online platform as a major part of their business. In addition, as is shown in Figure 8-1,

Table 8-1 List of public corporations by market capitalization, 2020 fourth quarter

Rank	Company	Market cap (million USD)	Sector
1	Apple Inc.	2,254,000	Electronics/Platform
2	Microsoft	1,682,000	Software/Platform
3	Amazon.com	1,634,000	Platform
4	Alphabet Inc.	1,185,000	Platform
5	Facebook, Inc.	776,590	Platform
6	Tencent	683,470	Platform
7	Tesla, Inc.	668,080	Automobile
8	Alibaba Group	628,650	Platform
9	TSMC	565,280	Semiconductor
10	Berkshire Hathaway	544,780	Finance

Source: Wikipedia and the author.

[1] Wikipedia. https://en.wikipedia.org/wiki/List_of_public_corporations_by_market_capitalization#cite_note-yba-17 (accessed on May 14, 2021).

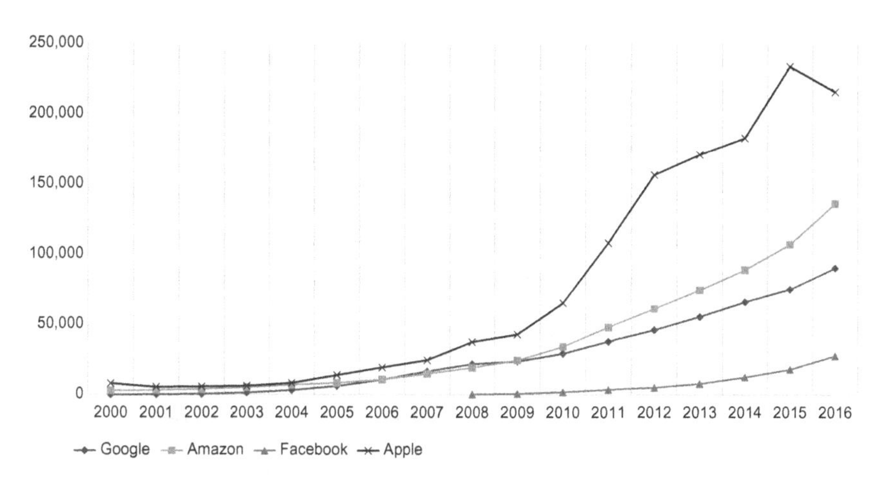

Figure 8-1 Sales of major platform companies

Source: Based on White Paper on Information and Communications in Japan, Ministry of Internal Affairs and Communications, Japan (2017).

sales trends of typical companies such as Google, Amazon, Facebook, and Apple, which provide online platforms, show rapid growth over just 10 years. All four companies, referred to by the acronym GAFA, are based in the US. In recent years, however, platform companies originating in China have also achieved rapid growth. Typical examples are Baidu (search engine), Alibaba (business transactions), Tencent (online communications), and Jindong (business transactions), which are referred to by the acronym BATJ.

Table 8-2 summarizes the top three companies in several countries regarding the usage status of platforms such as SNS.[2] The table shows that Facebook and YouTube have the highest share globally. Nonetheless, it should also be noted that, for different countries, country-specific platforms such as LINE, KakaoTalk, WhatsApp, Weibo, and WeChat have high utilization rates. Concerns have also been raised about the rapid growth of the global platform operators and their overwhelming share of users.[3] There are several issues in the mix, and one concern in terms of

[2]Ministry of Internal Affairs and Communications, Japan (2016). White Paper on Information and Communications in Japan. http://www.soumu.go.jp/johotsusintokei/whitepaper/ja/h28/html/nc132220.html.

[3]NIKKEI. https://www.nikkei.com/article/DGXMZO29883330W8A420C1MM8000/, http://www.meti.go.jp/meti_lib/report/2016fy/000179.pdf.

Table 8-2 Major SNS providers across countries

	No. 1	No. 2	No. 3
Japan	LINE	YouTube	Facebook
US	Facebook	YouTube	Twitter
UK	Facebook	YouTube	Twitter
Germany	Facebook	WhatsApp	YouTube
Republic of Korea	KakaoTalk	Facebook	YouTube
China	WeChat	Weibo	Renren
India	Facebook	WhatsApp	YouTube
Australia	Facebook	YouTube	Google+

pure industrial policy is that most countries rely on overseas companies in the platform field that can achieve high profitability. In June 2018, the "Future Investment Strategy 2018" compiled by the Japanese government stated that the goal is to "create 20 unlisted venture companies (unicorns) or listed venture companies with a corporate value or market capitalization of 1 billion dollars or more by 2023"; however, this shows that there is a concern that there are few world-class IT companies that have flourished in the domestic market.

The more a platform is used, the more valuable data about the user will be collected, which can be used as data for further mass customization and can in turn make the platform more competitive. Concerns have also been raised on whether, when resources such as data are gathered in a specific company, the company will seize a competitive advantage over the long term. There is also the idea that the accumulation of domestic personal information and security-related information by overseas platform operators may be an issue of privacy and national security. Due to these various concerns, there is an argument that some form of regulation is necessary for global platform operators.

However, from a competition policy perspective, just because some platform companies have a monopoly on competition, it does not mean that problems will arise immediately. It becomes an issue when a company makes use of this position and raises the price unreasonably and hinders new entrants, which distorts the market and results in some kind of disadvantage to consumers and users. It is at that time that appropriate interventions are required. Conversely, it is necessary to evaluate the type of information that should be protected regarding concerns about personal information and security. We must determine whether it is more secure to store data on a major platform or personally, also keeping convenience in mind. In any case, it is necessary to separate the issues of personal information, national security, and competition policy, and consider challenges and countermeasures.

Incidentally, can it be considered that the competition between platforms is settled, and that opportunities no longer exist for incumbent giant platform providers? Actually, there are a wide range of platforms apart from SNS and e-commerce. For example, GitHub, which Microsoft announced it would acquire for US$7.5 billion, is a platform used by software engineers worldwide for collaboration such as sharing source code and controlling versions. It is not widely known, as it is not used by general users, but it hosts the source code of cutting-edge software from all over the world and information on engineers who can develop the code, thus making it an extremely important platform in terms of technical knowledge. Microsoft's acquisition shows that they understand the importance of the technology development environment in the era of deframing, in which individuals play a leading role for innovation.

The platform Slack is more popular; it is similar to an SNS, specializing in work collaboration, and its use is spreading rapidly. It is simple to use and makes it easy to have discussions and share files. Additionally, as we saw in Table 8.2, there are specific platforms depending on the country, such as LINE, KakaoTalk, WhatsApp, Weibo, and WeChat, and it is not as though one common platform is used exclusively around the world. There is still plenty of scope to provide various platforms by field and by region.

Furthermore, there is also layering of platforms. One such example is the mini-program platform that operates within WeChat, a mobile application provided by Tencent. A mini-program can be developed by a third party and is similar to a smartphone app, but it is an "in-app app" that runs within WeChat. In that sense, it is a new layer of a platform that covers a smaller area than an smartphone platforms. Even though it is just an application, 1.2 billion users were using WeChat in the third quarter of 2020.[4] WeChat itself provides payment and communication functions, offering many mini-programs in combination.[5] In the past, apps were located on platforms such as iTunes for iOS and Google Play for Android devices;

[4]Tencent. https://mp.weixin.qq.com/s/Dq4p8njal5QUtoiQOjGl1w.

[5]https://lxr.co.jp/blog/7416/, http://www.catapultsuplex.com/entry/wechat-mini-programs-explained, https://36kr.jp/7008/.

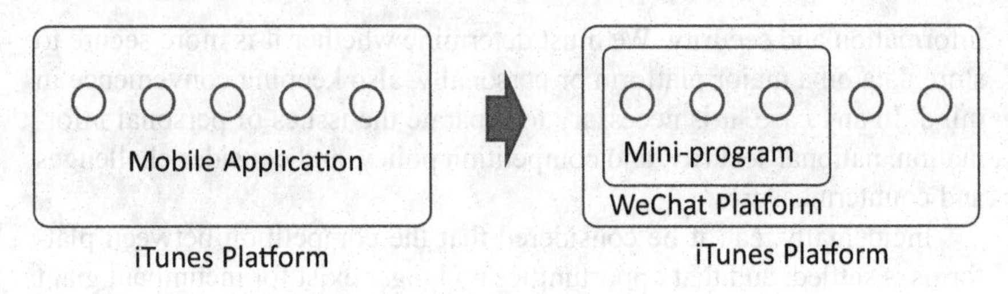

Figure 8-2 Layering of platforms

however, platform layering is spreading because the WeChat app functions as one platform (Figure 8-2).

One of such mini-programs is a game called "*Tiao yi tiao*" (roughly meaning jump jump). It is played by moving a figure with one's finger and is said to have over 300 million users cumulatively.[6] In addition, advertising in the game is also being considered, and there are reports that Nike and McDonald's are advertising on a trial basis.[7] Even the simplest games that are easily accessible to hundreds of millions of people with the power of digital technology can expand a huge business opportunity. The issue is understanding the scalability of the market when offering something.

In this way, there are still possibilities for platforms to be born out of various situations, and the idea of platforms still has the possibility for future growth. New platforms are being created every day around the world, and data are generated along with services. The amount of data is increasing exponentially every day, and rather than the idea of corralling that data, it makes more sense to think about how to create a new service that generates valuable data. There has been a growing discussion on regulating existing platform operators and trying to take back the control of data to avoid oligopoly, however, it is also important to consider how to create a service that produces data and how to locate ourselves to fit into the circle of such innovation.

Accumulation of Credit Information and Individual Independence

The second issue is related to the issue of platform monopolies and concerns the location where personal information is aggregated. As we move

[6]Diamond. https://diamond.jp/articles/-/219425.
[7]36kr. https://36kr.jp/7008/.

toward an age in which every individual is on the frontline, it will be important to evaluate each individual's past work and information that shows their credibility. Individuals have a harder time building a brand than a company, so they need to communicate their trustworthiness to potential trading partners. Therefore, there will be a need to have one's credit information verified.

Two types of businesses are in an advantageous position to provide this sort of personal credit information. One is a company that understands finance and settlement. If the degree of cashless payment increases and the digital analysis of information about personal income and expenses becomes easier, it will be possible to at least calculate financial credit. This is what Ant Financial, an affiliated company of Alibaba, is already doing with Zhima Credit. The other is a company that manages a social network and thus understands what kinds of people an individual has connections with. However, even if a person has many friends on SNS, it does not necessarily mean they have a high level of trustworthiness as a trading partners. With an analysis of their patterns and degrees of communication with others, it may be possible to somewhat calculate their trustworthiness.

Companies will want to understand the broadest range of information to get a richer and more accurate calculation of credit information. By collecting all kinds of information, such as employer, human networks, status on job orders received, and evaluation of that work, hobbies and eating habits, it is possible to provide an abundance of credit information. The more personal information is collected, the more sophisticated the analysis will be, and the company can act as a credit mediator to a wider audience through network effects. In the age of individualization brought about by the Internet, if credit mediation is the key, many Internet companies may plan to head in this direction.

Conversely, individuals may become more dependent on platform operators. The most dangerous situation would be if individuals who want their credit information to be rated higher become willing to provide even more information. As a result of the individualization of industries, it has been expected to possibly live with more freedom and independence, but in reality, it may end up that individuals become dependent on oligopolistic platforms and become desperate to raise their credit levels.

This aggregation of personal information and reliance on credit information will pose various challenges in the future. It is necessary to discuss whether it is appropriate for a large number of individuals to be dependent on a specific private credit information provider to conduct transactions. At the same time, the question is whether the government should be involved in this sort of credit information. In recent years, research has been conducted on ideas such as My Data and Information bank/personal data stores. My Data refers to the system that enables users of services to download the data containing users' personal profiles and history of usage of the services. It also enables the transfer of data to third-party vendors, so that consumers can easily switch providers, thus promoting fair competition in the market. On the other hand, an information bank enables users to deposit their personal data and let approved businesses to use the data, and the users can receive benefits from the business in exchange. Personal data stores are the mechanisms to mediate such data.[8] The Japanese government has issued a guideline for approving the information trust function in June 2018, and as of January 2021, six entities have been approved as providers to the information bank. These developments suggest that it is necessary to discuss points such as how people can control the information related to their credit and whether it is possible to ensure transparency regarding the calculation logic for credit information.

Furthermore, it is necessary to consider whether there is a risk that credit information will lead to long-term disparities. People with a higher credit rating because of their educational backgrounds and annual income may get more jobs, while those with lower credit ratings may not get work, which will further widen the gap. Excessive reliance on specific credit information also carries the risk of damaging an individual's independence. To avoid this situation, for work purposes, it is important to establish a community of direct human networks, without being overly digitally dependent. It is necessary to devise the environment so that there is a good balance.

Deframing and Privacy

The third issue is about privacy that is specific to deframing. Until now, when working as a company employee, one's personal information was protected by the company. This is because work is basically commissioned based on the trustworthiness of the company, and not designated

[8]Ministry of Internal Affairs and Communications, Japan. https://www.soumu.go.jp/main_content/000607546.pdf.

based on who is in charge of the work. However, in consulting work, a field in which employees have significantly different strengths, the client might request someone specifically. But only a limited number of clients have the tacit knowledge about who is the most reliable person in the company, and generally a transaction would start with the trust in the organization.

However, in the deframing age, there is no organization that will protect this kind of privacy. Individuals will have to bear the brunt of demonstrating their trustworthiness. Evaluations of the individual's work can be in the form of ratings and electronic word-of-mouth. It is possible that, among these, there could be incorrect information and unreasonable reviews. As seen in Chapter 6, reviews tend to be filled with extreme and negative opinions, which tends to deviate from the average impression of real users. When an individual is permanently exposed to unreasonable and extreme ratings on the Internet, this can cause the person to face difficulties in their future career. Therefore, in the deframing age, the management of information that is related to "decentralized trust" should be carefully conducted. One is the issue of information quality that platform operators will need to monitor and, in some cases, intervene. Some examples are unreasonably low ratings posted by competitors, posts created for reward, and so on.

Another is the issue of lifecycle management of information. Retaining information from many years ago indefinitely may not always be helpful for current transactions. In addition, the fact that past information is disclosed indefinitely is also an issue from the perspective of personal privacy. It may be possible to implement measures such as reducing the priority of reviews that were written before a certain period or deleting them altogether. As it is similar to the way the "right to be forgotten" is enforced in the EU, the design of the information life cycle is an important issue in the future, given that now is the time when content in the Internet remains indefinitely and when anybody can publish the opinion on whoever.

Need for Community and Social Security Mechanisms

Countermeasures for possible isolation are important in the era of deframing. As we have seen in individualization, we are in an age in which individuals will become important beyond the "frame" of companies. However, when working as an individual or working from home as a freelancer, one may go through the whole day without speaking to another person. Due to these circumstances, the number of "coworking spaces," as introduced in Chapter 5, is increasing rapidly. To enjoy work, rather than working in isolation, it is necessary to have real contact with a variety of people and have a sense of belonging to a community.

A research group led by Dr. Tuukka Toivonen (Central Saint Martins and University College London) suggests that the creativity of entrepreneurs is stimulated not just within a single space but rather realized by moving across multiple collaborative spaces.[9] They propose the concept of "networked creativity" based on an empirical study focused on tracing the behavior and movements of entrepreneurs in the London area. This suggests that individuals require not merely a single place or community, but rather a combination or network of freely accessible spaces and communities, for creative or innovative work in cities.

In Japan, the Freelance Association Japan is trying to build an open and flexible network of freelancers. For example, it is building initiatives for mutual aid such as building a community and skill and career enhancement support. It is also providing a freelancer database that lists the profiles of freelancing members so that other freelancers and potential clients can find proper partners.

On the other hand, countermeasure for income instability is also an important topic. As we saw in Chapter 7, the age at which a certain salary was guaranteed in a large company will gradually change to an age in which individuals are on the frontlines of work. In a sense, it suggests that this is an age of fierce competition. A survey by Freelance Association Japan reveals that the largest challenge of freelancing is unstable income, as 55.1% of respondents pointed out it is a challenge.[10]

On the other hand, corporate organizations will not necessarily disappear. In addition, it is each individual who makes the decision on whether to work in an organization or as a freelancer, and they may choose to work as an employee and a partly self-employed in parallel. However, the effects of deframing also extend to those who work in organizations. This

[9]Toivonen, T., O. Idoko, and C. Sorensen (2020). "An Invitation to the Unseen World of Networked Creativity: Tracing Idea Journeys through the New Infrastructures of Work." In *Collaborative spaces at work: Innovation, creativity & relations*, F. Montanari, D. R. Eikhof, E. Mattarelli, and A. Scapolan (eds.), Routledge, Abingdon, pp. 113–132.

[10]Freelance Association Japan (2020). Freelance White Paper. https://blog.freelance-jp.org/wp-content/uploads/2020/06/2020_0612_hakusho.pdf.

is because, if there is a freelancer outside the company who does the same job, and that freelancer is excellent and can offer a lower unit price, then the people inside the company will become embroiled in competition. People who work in an organization also need to more or less consider these factors of individualization.

In this environment, a major issue is how to replace the mutual aid function that corporate organizations have provided thus far. It is also necessary to develop a social system in which people can maintain a stable life, allowing for fluctuations in work and personal circumstances. Some insurance companies have begun to offer employment insurance for freelancers and professionals, and these are the seeds from which this kind of system will sprout. For example, the Freelance Association Japan is providing an insurance for freelancers that covers various risks such as the compensation associated with information leakage and defects in the delivery and decrease of income caused by sickness or injury.[11]

It is also necessary to create community functions, as seen in the story on coworking spaces, in addition to financial support. In the future, coworking spaces will not simply be shared office environments but are expected to grow into new spaces that respond to the individualized economy, with community functions in addition to those for shared spaces, employment insurance, and social insurance.

Meanwhile, considering the environment in which extended careers and individualization are progressing at the same time, providing support for re-learning is also important. Both the government and companies need to implement measures such as financial support for mid-career re-learning or "re-learning leave." In addition, as mentioned in Chapter 7, it is necessary to review educational curricula so that fundamentally, children learn the joy of learning from an early age and learn how to learn.

In the deframing age, individuals can develop their interests and strengths independently and act more freely. In addition to developing these positive aspects, it is necessary to concurrently implement institutional measures to maintain and enhance the stability of the society.

Restoring Cities as Cradles for Innovation

Just as coworking spaces are becoming more important as the frame of the company becomes more relaxed and individuals work more

---

[11] Freelance Association Japan. https://www.freelance-jp.org/benefits#insurance.

independently, cities themselves are becoming important as "cradles" for innovation. Until now, employees have spent most of their time in multi-divisional companies and have collaborated with each other here, so locations had only been considered in terms of factors such as proximity to customers and the abundance of the labor market.

However, the scale of business has become smaller, and in a society in which collaboration that goes beyond corporate boundaries is needed more than before, having a diverse number of people working in the same region and their ability to communicate easily has become more important. These kinds of structural changes have created areas that have become the birthplaces of new innovation.

One such example is Brooklyn, a suburb of New York. This was originally a rather less developed area, while Manhattan and New Jersey on the other side were relatively wealthy area. However, it was also for this reason that it was possible to live with low rent and for startups to move into Brooklyn. Due to the impact of this strategy to congregate startups in Brooklyn and because many coworking spaces were created, the area is growing today as a birthplace of cutting-edge culture and innovation.

Berlin, Germany, is also an area that is attracting attention. Some areas were devastated by the division of East and West Berlin for many years, but by renovating factories that were abandoned after the unification and leaving the historic areas as they were, Berlin has grown rapidly as an area where many artists and startup companies gather.[12] By gathering diverse people who place importance on sensitivity, a movement may start that will question social common sense and realize new values and experiences. These new values could lead to new services.

The east side of London, the area known as the East End, has also undergone similar developments. It was not a very wealthy area to begin with, which is why artists and entrepreneurs moved there in search of cheap rent. Artists need a large space to carry out their creative activities, and warehouses and factories that are no longer in use are just the right places for this. Since about 2006–2008, East London, especially areas such as Hackney Wick and Fish Island, where the main venue for the 2012 London Olympics was located, has become a gathering place for those who work in creative fields and is becoming a center of innovation.

---

[12]Takemura, M. (2018). *Berlin, City, Future* (original in Japanese, trans. by author), Ohta Shuppan, Tokyo.

In all these cities, human resources rich in diversity and creativity, such as artists and entrepreneurs, gather in relatively poor areas along the outskirts due to the low costs. These people inspire one another, and the areas grew as centers of innovation. That said, this does not mean that a poor area will see growth if it is left as-is. This type of growth process can only be achieved by combining a living environment that stimulates creativity, designing the space to encourage comfortable interaction, having low living costs, and attracting key artists and entrepreneurs.

The idea of fundamentally transforming social structures and business models is attracting attention under the phrase "digital transformation," but to that end, we must question conventional wisdom and need to conceive completely new values and social structures. This way of thinking is deeply connected to art and social movements. To promote innovation and business in the region, these kinds of diverse human resources must gather, and thus it is necessary to consider the development of a city from a much broader perspective than the conventional method of regional and industrial promotion.

Of course, if the region grows too quickly, there is a risk that rents will rise and the people who were the leaders of innovation will move away. This is a phenomenon called "gentrification," which is occurring in San Francisco and is becoming an obstacle to Silicon Valley's future as an innovation center. Growing while maintaining an area of diversity and creativity is an important issue in the regional management and in promoting innovation.

Afterword

As discussed in this book, innovations in information technology are reintegrating operations, transforming business models, and even affecting workstyles and organizational structures at an unprecedented speed.

This book is a continuation of my book, *Reweaving the Economy: How IT Affects the Borders of Country and Organization* (University of Tokyo Press, 2017). This academic book dealt mainly with topics on the economic impact of offshoring of software development and call centers, and also of cloud computing. I also touched on collaborations between individuals, however, the main topic was the outsourcing of business units undertaken by companies. Even at that time, the foundation was the reduction of transaction costs through information technology, and there were indications that eventually, transactions would be carried out between individuals and not only between businesses.

Years have passed since then, and in reality, the presence of individuals in the economy increases each day, the specifics of which are described in this book. I was also involved in blockchain research, which led me to meet many young entrepreneurs in this space. In addition, the career paths of the graduates of The University of Tokyo, to which I belong, are not limited to becoming public officials or large corporate employees but are extended to joining startup companies, starting their own businesses, or working in small, highly specialized companies.

Over the past several years, I have personally experienced the effects of deframing. At the Business Breakthrough University, I was involved as an individual in the creation of blockchain-related educational content, and I also became acquainted with the examination of educational content to be distributed on the Internet.

Those who have stable employment in large companies may feel disassociated from the trend of individualization. Of course, long and stable employment is valuable. However, as shown in this book, there is no guarantee that a person will be protected forever in the frame of a company, in an era of the super-aging society and when the limits to the sustainability of social security are apparent. It is valuable to receive the benefits of stable employment, but it is also necessary to prepare for the future.

Deframing can be seen in both light and shadow. On the one hand, it allows the potential of new services, new technology, and individual creativity to be fully maximized. On the other hand, it can make the economy unstable. Making the best use of the light areas and compensating for the shadows is an issue to address over the next few decades.

I hope that the content presented in this book will give you an opportunity to think about how to respond to these considerations in the future.

Index